GEOCHRONOLOGY IN CANADA

Edited by

F. FITZ OSBORNE

This volume is made up of papers presented at a colloquium of the Geology Division of Section III of the Royal Society of Canada at the annual meeting in Quebec, June 1963. The papers fall into two groups: in one group the validity and shortcomings of the methods of establishing the geological time-table are discussed; and, in the other, applications of the methods to areas across Canada, and from Precambrian to recent, are described.

The geological time-table has been built up from the record of the rocks and is based on the law of superposition, a fact that is pointed out in the first paper of this volume. The chronological value of fossils, palaeomagnetism as a means of dating geological events, the limitations of radiometric dating, and other pertinent matters are here dealt with by a group of wellknown authorities. These scientific disquisitions will be of great importance to geologists everywhere.

This work should be of special interest to those engaged in research on the history of the earth, particularly in relation to the nature, the causes, and the time of an event. It will also serve as a valuable reference to practising geologists in government or industry, to university departments of geology, and to geological consultants.

The distinguished contributors include members of several Canadian university departments of geology, government departments, and geological consultants. The editor, F. Fitz Osborne, F.R.S.C., is Professor of Petrology, Laval University.

THE ROYAL SOCIETY OF CANADA

Special Publications

GEOCHRONOLOGY
IN CANADA

THE ROYAL SOCIETY OF CANADA
SPECIAL PUBLICATIONS, NO. 8

Edited by F. Fitz Osborne

PUBLISHED BY THE UNIVERSITY OF TORONTO PRESS
IN CO-OPERATION WITH
THE ROYAL SOCIETY OF CANADA
1964

PREFACE

THIS VOLUME is made up of papers presented at a colloquium of the Geology Division of Section III of the Royal Society of Canada at the Annual Meeting in Quebec, June 1963. C. S. Lord acted as chairman of the colloquium, and C. H. Stockwell arranged for the papers composing it.

The papers of the volume fall into two groups: in one group the validity and shortcomings of the methods of establishing the geological time-table are discussed, and, in the other, applications of the methods to areas across Canada, and from Precambrian to recent, are described.

The editor is pleased to express his appreciation to Miss L. Ourom, Assistant Editor, University of Toronto Press, for her services in preparing this volume.

F. F. O.

CONTENTS

CONTRIBUTORS

H. BAADSGAARD, *Department of Geology, University of Alberta, Edmonton, Alberta*

G. L. CUMMING, *Department of Physics, University of Alberta, Edmonton, Alberta*

R. DE WIT, *J. C. Sproule and Associates Ltd., Calgary, Alberta*

D. L. DINELEY, *Department of Geology, University of Ottawa, Ottawa, Ontario*

A. DREIMANIS, *Department of Geology, The University of Western Ontario, London, Ontario*

R. E. FOLINSBEE, *Department of Geology, University of Alberta, Edmonton, Alberta*

H. GABRIELSE, *Geological Survey of Canada, Ottawa, Ontario*

J. D. GODFREY, *Geology Division, Research Council of Alberta, Edmonton, Alberta*

D. G. KELLEY, *Geological Survey of Canada, Ottawa, Ontario*

A. LAROCHELLE, *Geological Survey of Canada, Ottawa, Ontario*

L. W. MORLEY, *Geological Survey of Canada, Ottawa, Ontario*

E. R. W. NEALE, *Geological Survey of Canada, Ottawa, Ontario*

F. K. NORTH, *Department of Geology, Carleton University, Ottawa, Ontario*

W. H. POOLE, *Geological Survey of Canada, Ottawa, Ontario*

J. E. REESOR, *Geological Survey of Canada, Ottawa, Ontario*

C. H. STOCKWELL, *Geological Survey of Canada, Ottawa, Ontario*

GEOCHRONOLOGY IN CANADA

INTRODUCTION

F. Fitz Osborne, F.R.S.C.

THE DISCIPLINES that make up geology are united in being concerned with the history of the earth. In general, the ends of the science are satisfied if the nature, the causes, and the time of an event are determined. The events can be of greatly different order of magnitude: the appearance of a species of microfossil can be as significant to some palaeontologists as the development of a range of mountains is to some geologists. Time can be considered as the string that holds together and in order the beads of geological events. Geologists are not the sole proprietors of time; however, they exercise a right, in much the same way as do the historians in referring to time spans by names such as Dark Ages and Middle Ages, to use names to refer to time spans. The principal named divisions constitute the geological time-table, concerning which there is general agreement among geologists.

The geological time-table has been built up from the record of the rocks and is based on the law of superposition, a fact that is pointed out in the first paper of this volume. However, this law must be considered as embracing more phenomena than is suggested by its name: intrusive contacts and structural relationships can and must be taken into consideration. In general, only the relative ages of two adjacent beds or formations can be established. Nevertheless, by painstaking use of the law of superposition, a sequence of rock units has been established, and names have been given to parts of the sequence. The names are considered to indicate time in the geological time-table.

The use of fossils proved of very great assistance in applying the time-table to Phanerozoic rocks. In view of the fact that index fossils are to be found in beds that have a position in the sequence of rocks, the time-table as established from the rocks is definitive. Discrepancies are attributed to errors in observation or deduction on the part of either the geologist or palaeontologist. The study of palaeomagnetism provides a method of dating rocks that is exactly analogous to that using fossils: both depend on comparisons with observations made on rocks that have been otherwise dated. The ages arrived at by these methods are those of the deposition of a sedimentary rock or of crystallization of igneous rocks, and these are, conventionally, considered the geological ages of the rocks.

Yet another method of establishing dates has come into use. This method is based on the ratio of a daughter element to the unstable parent element contained within a mineral or rock. Because the rate of formation of the daughter element can be determined in the laboratory, the calculated age is not dependent on the time-table established by the law of superposition. However, the isotopic ages do not give a second time-table. To refer to a metaphor used earlier in this introduction, the isotopic ages can show the spacing of the beads on their string, and under favourable conditions the size of the beads and the length of the string. However, the isotopic dates can represent the time of a process that acted on the beads after they were strung. The fact that the isotopic ages refer to the results of a process operating after the formation of a rock means simply that the isotopic ages do not necessarily refer to the age of the rock. These ages provide data leading not to a second time-scale but perhaps to an arrangement of events other than the time of formation of the rock.

The nature of the events that determine the dates fixed from the ratios of isotopes has not been established fully. Many dates are considered to correspond to an orogeny and this is a conclusion consistent with the contention, now somewhat outmoded, that orogenic events tend to separate major units in the time-table. The papers on the Appalachian and Cordilleran regions illustrate the way in which isotopic dates are used to amplify the geological history of areas. The way in which the isotopic dates for minerals from the Canadian Shield are to be interpreted is unfortunately less clear than for those from the Appalachian and Cordilleran regions. There can be little doubt that the isotopic dates correspond to an event or events that occurred within the time span indicated by the limits of error for the individual determinations, but the time span is long, even exceeding that of a Phanerozoic era. With the large number of dates available from minerals, particularly biotite, from the Canadian Shield, it is safe to assert that very large areas of the region have dates of the same order of magnitude, but dates which are different from those of adjacent large areas. These large areas have been called tectonic provinces, but the great extent of each and the relationship between them, as well as other dissimilarities to Phanerozoic tectonic provinces, suggest that some other process or processes in addition to or apart from tectonism played a role in impressing the dates on the minerals.

THE GEOLOGICAL TIME-SCALE

F. K. North

ABSTRACT

The skills of the palaeontologist and the radiometric dating specialist serve ends as different as their methods. They can never be in insoluble conflict. The palaeontological time-scale, founded on the law of superposition and independent of any assumption, is not a scale at all, but a succession without dimensions. The radiometric scale is a series of charted control points along a succession much of which is already established.

Specialists in both fields need constantly to remind themselves that they are not operating exact sciences. Both make serious errors, and when their decisions seem in conflict the ring is still held by the geologist on the ground.

A GLANCE through the abstracts of the articles included in this volume might lead one to the conclusion that two geological time-scales are now in existence, one founded on the development of organic, the other of inorganic, materials. There can, of course, be only one true time-scale in a sense having any meaning for most of us. What is certainly true, however, is that the scale which has for so long been built up patiently by a single means is now rapidly being fortified and elucidated by another one. And this newer tool differs so radically from the older one, both in its means and in its ends, that it may cost us some effort to bear in mind that the two must eventually contribute to the same edifice and not to two quite different ones.

The palaeontological time-scale rests squarely on the law of superposition, independent of any theory or any assumption. From this unassailable foundation, the palaeontologist became for more than a century the arbiter of all stratigraphic organization. But, for geologists, the law of superposition presupposes the existence of decipherable geological sections, and every geological section must have a top and a base. The palaeontological succession was pieced together from hundreds of such sections, the tops and bases of which had been established by geologists on the ground. The palaeontologist's wheel of authority turned full circle when he put this process into reverse, and used his fossils to determine tops and bottoms for himself. In the course of time he came to rule upon stratigraphic order, and on gaps within it, on a world-wide basis.

The succession the palaeontologist produced is now essentially complete, and a very wonderful piece of work it is. Some difficulties about systemic and serial boundaries remain. But these are difficulties about where to draw the boundaries, not about whether sections truly crossing the boundaries

have been found. Eventually, the difficulties will be sorted out by agreement. Oddly enough, it is the very latest of the boundaries, that between
the Pliocene and the Pleistocene, that may persist in uncertainty the
longest, and in whose determination from place to place the techniques of
the palaeontologist may be least helpful, and those of the radiometrist
decisive.

The stratigraphic palaeontologist has, of course, always known that his
trusted, and trustworthy, system breaks down in the metamorphic regions
of the world, especially in the ancient shields. But it does not break down
primarily because of the lack of fossils. It breaks down because of the lack
of continuously correlatable sections. It breaks down, in fact, where the
law of superposition loses its meaning. It loses it in Mesozoic ophiolitic
belts almost as completely as it loses it in the shields. At the same time, it
is entirely valid in many sections of totally unfossiliferous rocks. It is not on
the basis of fossils (though they contain some) that the formations of the
Belt and Purcell series are confidently correlated, or the Navajo sandstone
or the Columbia River basalts traced without uncertainty through scores of
widely separated outcrops.

It is where the law of superposition loses its usefulness that the radiometrist has become king. The great development of his techniques for
dating rocks is encouraging some people to conclude that the new methods
and the old one are rivals. If they are rivals, they are rivals only in the
sense that all the sciences are in rivalry. They are not rivals for the same
prize. The palaeontologist makes no claim for the dimensions of his succession. He has produced order without scale, and, moreover, order over only
one-sixth or less of the stratigraphic record. Within this one-sixth, radiometric dating merely proposes the dimensions, calibrates the scale. It also,
happily, proposes the scale over the other five-sixths of the record, for
which the palaeontologist is silent (or nearly so).

Over the one-sixth of concern to palaeontology, well-founded palaeontological data can never be contradicted by radiometric data, and even when
the data are in error the errors are likely to be on a smaller scale than the
margin of error of most radiometric dates, which are, after all, based on
more than one assumption. For example, radiometric dating in the world's
Precambrian shields is offering support to the idea of world-wide synchroneity, in a generalized way, of major orogenic episodes. Yet the spread of
dates for which a mean is taken to "typify" a Precambrian orogeny represents the duration of a whole geological era, whereas individual pulses in
the post-Cambrian orogenies have been pinned down accurately by fossil
evidence to within a fraction of a geological period, perhaps (as in the
cases of the late Palaeozoic deformations in the Ouachita belt, and the
Cretaceous upheaval of the Cordillera) to within a geological age.

The danger is that practitioners of the two disciplines tend to fall into
the same two errors, quite apart from that of imagining themselves to be
in competition. Each suffers from incomplete appreciation of the other's

methods, and of both the other's and his own limitations. And each easily acquires the habit of believing (though he would never claim to believe) that he is operating an exact science.

I said that well-founded palaeontological data can never be contradicted by radiometric data. It is the ill-founded palaeontological data, the palaeontologist's errors, which confound us. As we are discussing the geological time-scale, I shall restrict my examples of ill-founded palaeontological data to two of the groups of organisms on which stratigraphic dating is based: ammonites and foraminifera.

A few errors stem from an incorrect understanding of the actual order of superposition. An intriguing recent example comes from Dr. Raymond Casey, in Britain, who has revealed that a well-known Lower Cretaceous zonal ammonite is actually older than an equally-well-known Upper Jurassic zonal ammonite and, where the two occur in one section, the former is underneath.

Other errors are due to premature certainty concerning the real range of a guide fossil or fauna. Paradoxical stratigraphic relations in the Jurassic basins of western Argentina and Chile, for example, seem to spring from the assumption, by oil company palaeontologists, that ammonite *genera* found in them have exactly the same stratigraphic range in South America as they are said to have in Madagascar, where Collignon described them.

In younger formations, the chronostratigraphy of which is based on foraminifera, the danger of reworked faunas is often overlooked. Foraminifera having the same size as sand grains, and approximately the same shape, have bedevilled a lot of correlations in the deep basins of Tertiary tip-heap sedimentation.

Commonest of all is the human error of misidentification, which we dare not hope will ever be eliminated. Except in standard localities, more fossils are identified wrongly than rightly. Those of you who are familiar with the late W. J. Arkell's great compilation on Jurassic geology will recall how many suites of ammonites from vital localities he dismissed as wrongly identified. Can we be certain his own reidentifications are correct? Or that anybody else's would be?

Along the north coast of Venezuela lies a mountainous belt of metamorphosed and semi-metamorphosed sediments and volcanics, with the city of Caracas in the middle of it. On regional grounds, the rocks of this belt ought to be pre-Aptian; they are very probably Paleozoic. Some fifteen years ago, a well-known palaeontologist described a group of foraminifera from a locality in the very heart of the belt; he identified them at specific level, and pinned the beds yielding them to a single Upper Cretaceous stage, the Turonian. This identification had far-reaching effects, because it was one of the factors which influenced a very famous geologist indeed to assign all the rocks of the deformed belt to the Cretaceous, and so to throw a gigantic wrench into the mill of Caribbean geology (which can cause almost as much despair as Caribbean politics). The growing suspicion that this assignment could not be correct eventually led the same palaeontologist to restudy his material. The outcome is a second paper, withdrawing the identifications, and admitting that the bodies are not foraminifera, that they may not even be fossils, and that if they are fossils the author does not know to which phylum they belong. Let us beware.

The scale of potential radiometric errors is demonstrated by the current treatment of the stratigraphic periods as if they were the folds of an accordion, to be stretched or shortened at will. It is apparent that not all the controls required for the methods are yet perfectly understood (the rate and manner of argon loss, for example). No one, of course, pretends that they are, but the results of the measurements are expressed in figures, complete with margins of error, and we know the aura of authority that seems to surround anything that looks quantitative.

Genuine conflicts between radiometric dates and palaeontological dates are usually due to dubious interpretations of the geology: of intrusive relations, notably, or of periods of recrystallization of the minerals used by the radiometrist. I remember vividly the astonishment of a party of geologists (actually in Cuba, and myself among them), who had mapped a body of granodiorite which all agreed was unconformably overlapped by a fossiliferous Portlandian limestone, when they later learned that their carefully collected mica samples had been radiometrically dated, not as Nevadan (as they feared), still less as Hercynian (as they hoped), but as Eocene, by the men of Lamont. Perhaps, therefore, I should admit to some qualms in offering a final suggestion: that, in cases in which the palaeontologist and the radiometrist are really in conflict, the ring will have to be held by that otherwise old-fashioned looker-on, the geologist in the field. Meanwhile, his services will continue to be available to them, and he wishes them well.

THE CHRONOLOGICAL VALUE OF FOSSILS

D. L. Dineley

ABSTRACT

The principle that faunal and floral succession can be used to determine the relative ages of strata was recognized by William Smith and has since been applied to correlation, though facies fossils and homeomorphs have caused problems. Relict faunas occur but do not support Huxley's idea of homotaxis. Biological evolutionary processes are very rapid compared to most geological processes influencing the stratigraphic record. There are differing concepts of how, in detail, biostratigraphic principles should be applied but all are based on the simple fact that the palaeontological succession, tied to guaranteed stratigraphic sections, remains the geologist's surest means of recognizing relative age.

EVERY GEOLOGY STUDENT is soon familiar with the concept generally known as the order of superposition, and its importance has been recognized since the days of Lehman and Werner. We may be reasonably sure that Steno, writing in Italy in the 17th century, appreciated it. With the work of William Smith in Somerset in the west of England towards the end of the 18th century, however, the principle was not only demonstrated to a scientific public but also it was shown that geological formations can be identified by the fossils they contain. Fossils characteristic of distinct rock formations were seen to identify the geological times in which the strata were formed. Smith was not the only one to make such a demonstration: the Abbé Giraud-Soulavie in 1780 divided the strata of Vivrais in southern France into five groups on the basis of their fossils. The fossils seemed to many of the early stratigraphers to be intimate parts of the formations, perhaps rather like trade marks, serving to identify rock layers. In a comparatively short time the stratigraphical column included most of the systems now recognized; and it was essentially a rock-unit succession. Fossils helped rock correlation, and, as further collections were made, a succession of faunas and floras independent of lithologies was recognized. The highly fossiliferous Jurassic rocks of England and France lent themselves particularly well to the early stratigraphical studies. It is not surprising that great strides in palaeontological correlation and identification of strata were soon to be made in these rocks by Hunton in Yorkshire and by D'Orbigny in France. D'Orbigny made rapid progress and divided the Jurassic strata into 10 stages "solely according to the identity in the composition of the faunas, or the extinction of genera or families" (Arkell 1933, pp. 9–10). Extending his work beyond France, D'Orbigny became convinced that his stages were uniform isochronous entities that could be found across the whole earth. The German geologist Quenstedt was able to offer a telling criticism of D'Orbigny's later views and demanded that stratigraphic schemes be tied to exactly described geological sections.

Again in Germany, a further subdivision of stratal units by reference to fossils was achieved by Oppel, who was the initiator of zones determined by the vertical range of species—"zones which, through the constant and exclusive occurrence of certain species, mark themselves off from their neighbours as distinct horizons." Since then such an amount of ink has been expended upon the zonal concept that its essential simplicity often seems lost. Among the best historical accounts is still that of Arkell (1933), and to this the reader is referred for details. Wheeler (1958) gives a good discussion of the terminology of zones. The concern of most of the pioneer stratigraphers was to establish a succession. While their series, stages, and zones are essentially tangible entities, the realization that these represent diverse intervals of geological time was also there.

During the last sixty years or more, an immense amount of work has been done in subdividing the stratigraphical column into smaller and smaller units and in producing with this a more detailed chronology. Stages and zones are established in every system, and in many different facies; and each is regarded as being essentially isochronous throughout its extent. To what extent is this belief justified?

The palaeontological record is still lamentably incomplete even though a detailed succession of forms from Cambrian to present times is known and a great deal has been added to our knowledge of organic evolution from its study. The very fact that evolution takes place is largely responsible for the success of palaeontological dating of the rocks. It provides an excellent and universal *relative* time-scale, but not an *absolute* time-scale. Almost every phylum has been pressed to stratigraphical use in one part of the column or another (see Teichert 1958), and rates of evolution in different groups within a phylum may vary very widely. Various estimates have been made of the speed at which certain evolutionary changes appear and spread throughout populations, and again much of this work establishes a *relative*, not an *absolute* scale. Foremost among contributors in this field has been G. G. Simpson (see especially Simpson 1944), while F. E. Zeuner (1946) has discussed organic evolution and geochronology at some length and with interesting conclusions.

Geochronology today is largely in the hands of the mineralogist or radiochemist, but it is interesting to recall that many years ago Charles Lyell (1867) made a remarkable estimate of Tertiary and Palaeozoic time on the basis of his studies of molluscan faunas. He suggested that Pleistocene and Recent evolution is not more than one-twentieth of that which has taken place since the beginning of the Miocene. Since Miocene times all the species existing then have been replaced, this being what Lyell regarded as a "cycle of evolution." Lyell accepted Croll's figure of one million years for the Pleistocene and Recent and deduced that the base of the Miocene must be 20 million years old. With four "cycles" in the Tertiary era, some 80 million years are indicated, and with 12 "cycles" since the beginning of the Palaeozoic a figure is given of 240 million years since then. Most palaeontologists would today be sceptical about the number and nature of the "cycles"—many more than 12 since early Palaeozoic days might be indicated. Lyell's method may not have great appeal today, but his estimates are surprisingly good. Geologists generally are content to use fossils for *stratigraphical* purposes without attaching any significance to them from an *absolute geochronological* viewpoint.

From the discussion that follows it will be apparent that, although there are some very real problems that the stratigraphical palaeontologist has to face when drawing up his time-scale, the magnitude of time involved for the organic changes he utilizes in fossils for chronological purposes is of a different order from that required for the processes of rock accumulation.

CORRELATION, ZONES

The fundamental method of stratigraphy remains, as advocated by Quenstedt, the careful collecting and identification of fossils contained in stratigraphical units. Similar assemblages collected from successions in different localities have always been taken to indicate an approximate time-equivalence. Stratigraphic correlation still depends mainly upon this method, though it is beset with many difficulties which are even now not always fully appreciated by palaeontologists. Facies changes, migration, biological provinces and basins of deposition, changing rates of evolution, and incomplete successions are all hazards attendant upon correlation by fossils and to these must be added the still inadequate state of comparative palaeontological research.

While the early stratigraphers had little difficulty in indentifying large groups of strata by the "fossil biota," it has become obviously less easy to make certain correlation as strata are divided into smaller and smaller units. The reasons for this are fairly obvious. Floras and faunas in successive units, perhaps formed under restricted depositional conditions, may resemble closely others of different age. Such difficulties have been encountered on different scales, for example, in both the Carboniferous and Pleistocene strata. A further obvious difficulty is that population variations in synchronous deposits may be greater than differences between contiguous assemblages. Nevertheless, palaeontological correlation produces a succession in which it is held that the same fauna is everywhere of essentially the same *geological* age. The palaeontologist is concerned to know more about the fauna itself while the stratigrapher is interested in its use to identify a significant unit of geological time. Absolute age rarely concerns them, but relative age is invariably of major importance. The zone is usually considered as the basic (perhaps one should say the *ultimate*) unit to be identified, representing the smallest span of *identifiable* geological time in our scheme.

Zones are generally accepted as units of strata, each characterized by an assemblage of fossils, one of which is taken as the index form. The term is also encountered in the words biozone, faunizone, teilzone, and so on, each of which denotes a significant unit, the definition of which is not always apparent from the way different authors have used it. Often also the word "zone" has been used loosely where use of one of the other terms would avoid confusion. Basically, however, all are based upon the field occurrence of fossils—their ranges, abundance, and associations. Sometimes it is the *total* assemblage of fossils that affords the distinctive character of a zone; other zones are recognized by the state or *stage of evolution* of one of the component species.

The ideal zone fossil has a wide geographical distribution and a short stratigraphical or vertical range (i.e. the species was evolutionarily short-lived) and, of course, it should be reasonably common. Clearly there is no uniform thickness to a zone—the actual amount of sediment present is of minor importance. From one locality to another a zone may visibly change its thickness in a spectacular way, as seen in many geosynclinal regions. An interesting instance of such a change was also demonstrated in the "Shannon Trough" of the Namurian (Carboniferous) of Eire by Hodson and Lewarne (1961). Local events following or even during deposition may interrupt the succession. Many breaks are immediately apparent as disconformities with attendant obvious lithological features; others are diastems, which escape the eye, as for example those demonstrated in the Oxfordian of England by Brinkmann (1929).

As stated, almost every type of fossil may be used in correlation to a degree. Even trace fossils may be utilized to this end in rocks of quite different—even Precambrian—ages. Although macrofossils have been the means by which most correlation has been effected so far, micropalaeontology has made great strides in the last 30 years, with foraminifera, ostracods, and, more recently, conodonts having been employed most successfully. Marine invertebrates dominate the field but there is an increasing use of other groups, such as freshwater fish, mammals, and even insects. A flourishing branch of palaeontology, most useful to the stratigrapher, has appeared in the relatively new palynology. By far the best correlating aids seem to be the marine animals of planktonic habit, possessed of robust skeletons with rapidly evolving morphology. The ammonoids are the obvious and well-known example, but certain foraminifera fill the role equally well (in different rocks). Much of the earliest work on zones (in the Jurassic of Europe)—and it was based on ammonites—has stood the test of time very well. The epiplanktonic graptolites, though less susceptible to preservation, proved in Lapworth's pioneer work to be extremely valuable. He found them widely distributed, many evolving rapidly and restricted to small thicknesses of rock. Though now somewhat modified, his work on the Ordovician and Silurian graptolite successions has served well in widely separated parts of the world. Not all the units are equally wide-spread, some having been found in only a few places.

Most series of rocks contain more than one group of stratigraphically useful fossils, though the vertical ranges of these are unlikely to coincide. Rarely can stratigraphical units be defined in the same zonal terms by two different groups of fossils. This is not to say that different groups cannot be employed together to produce a workable scheme. Experience so far seems to indicate that a scheme founded upon a single group of animals and, if possible, on a single evolutionary lineage within the group provides the best means of zoning.

As examples of schemes wherein the criteria are *assemblages* of fossils we may quote the classic work of Vaughan (1905) in the Lower Carboniferous of Britain with its coral-brachiopod assemblages (now regarded as a local facies fauna by some), and the European chalk. The latter is divided as follows:

Upper Senonian	*Belemnitella lanceolata* or *Ostraea lunata* *Belemnitella mucronata* *Gonioteuthis (Actinocamax) quadrata*	Belemnites
	Offaster pillula	Echinoid
Lower Senonian	*Marsupites testudinarius* *Uintacrinus westphalicus*	Crinoids
	Micraster coranguinum *Micraster cortestudinarium*	Echinoids
Turonian	*Holaster planus* *Terebratulina lata* *Inoceramus labiatus*	Echinoid Brachiopod Pelecypod
Cenomanian	*Holaster subglobosus* *Schloenbachia varians*	Echinoid Ammonite

Of the fossils listed, the *Micrasters* and very possibly the *Belemnitellas* are parts of recognizable evolutionary series. Work in Europe now suggests that the pelecypods of the *Inoceramus* group may be used in a zonal scheme. Foraminifera evolved rapidly during the Upper Cretaceous, and zonation by these fossils and by coccoliths is now possible.

Each of the different groups of fossils mentioned has evolved at its own rate and sometimes at widely differing rates throughout its geological life. Zones based upon a group at one stage or phase of its evolutionary life might very well be of a different time span from zones taken at another phase. We may ask how long it takes to produce a zone, and the answer depends very much upon the type of zone (foraminiferal, coral, mammalian, etc.) chosen.

The Ordovician period is regarded as having lasted for 75 million years and is marked by 20 graptolite zones in North America. Hence, it might be assumed that the *average* length of a zone here is a little over 3½ million years. On a similar basis the average length of Silurian graptolite zones may, with 21 zones in the system, be about 1 million years. Jurassic ammonite zones have been estimated at 300,000 years or less in duration.

LOCAL SPECIES AND HOMEOMORPHS

There are certain features which sometimes diminish the obvious advantages in using for correlation groups subject to rapid evolution. Of these, perhaps the most familiar are (1) the tendency to undergo local evolution and hence to produce local instead of wide-spread species and (2) the tendency to produce homeomorphs.

As an example of the first the Lower Devonian pteraspids (ostracoderms) in Britain show local variations which appear so far to be limited in both geographical as well as stratigraphical extent. Although the coeval pteraspid successions in other parts of western Europe, in Spitsbergen, and in North America show similarities, the known actual spread of British variations, and even species, is almost nil. Usually it is the case that such fossils are "facies fossils" (see below) or are in part of a succession laid down under peculiar conditions, usually non-marine or at least not normal marine conditions.

Homeomorphs offer more trouble, but even this is resolved in known areas or where they are from deposits of different ages. Ammonites are again the classic examples of this (see Arkell *et al.* 1957). The well-known Lower Jurassic (Liassic) genera *Coroniceras* and *Paltechioceras* produce close homeomorphs, but they are five zones apart and seldom confuse European collectors. Where heterochronous homeomorphs occur in geologically poorly known areas, however, confusion may result. When the palaeontologist is concerned with *partial-genera* or *form-genera*, as in the case of conodonts, some vertebrates, etc., homeomorphs may be common in any succession. But this has not led to any stratigraphical "howlers" so far as I know.

FACIES FOSSILS

Facies variation in a group of strata denotes lateral changes in conditions of environment and deposition. The local biota was subject to the influence of these changes. Temperature, salinity, depth of water, turbidity, food supply, and bottom conditions are such factors controlling marine communities. Although forms pelagic throughout or during part of their lives are normally widely distributed by marine currents, sessile benthonic forms are often more restricted in the type of sediment in which their fossils are found. Most fossil coelenterates—the rugosa are noteworthy examples—are restricted in their mode of occurrence. Such preference for a type of habitat and—eventually—sediment type makes a fossil a facies fossil to a greater or lesser degree. It might be argued that all fossils are facies fossils to some extent, but the implication in the general use of the term is that this form is not found in all coeval strata but only in those of a particular facies.

Reefal developments in epeiric shelf sequences offer peculiar and sharply defined facies in which the fossil assemblages are remarkably different from those in the contemporaneous open water and lagoonal deposits. Such are well known in the Silurian of the Michigan Basin (see Lowenstam 1950), the Welsh Borderland, and Gotland. The essential members of the reefs are stromatoporoids, algae, and tabulate corals, rarely found in the local non-reefal strata. Trilobites, brachiopods, bryozoa, crinoids, molluscs, and other fossils found with the reef builders are usually quite distinct from those of the other nearby environments. Correlation here has often been very difficult. Indeed, the correlation of the Lower Carboniferous reef knoll limestones with the "shelf" and the "basin" deposits in northern England was the subject of dispute for several decades (see Bond 1950). It has often been the case that correlation between adjacent different facies has been determined more immediately from the field relations than from the fossils.

Facies fossils, then, clearly are of importance as indicators of habitats and may be used stratigraphically *only* within their own facies. Their use is thus strictly limited.

Homotaxis

Study of the distribution and migration of modern or at least Recent forms of life led several critics of Lyell's stratigraphical and palaeontological concepts to suggest that identical fossils found in distant areas could not be of the same age because of the time required for migration. T. H. Huxley was among these critics, pointing out that similar faunal successions in different regions must be of different ages. He proposed the term *homotaxis* for this similarity. In fact just over 100 years ago (1862) he said that there seems to be

no escape from the admission that neither physical geology nor palaeontology possesses any method by which the absolute synchronism of two strata can be demonstrated. All that geology can prove is local order of succession . . . but the moment the geologist has to deal with large areas or with completely separated deposits, then the mischief of confounding that "homotaxis" or similarity of arrangement which can be demonstrated with "synchrony" or "identity of date," for which there is not a shadow of proof under the one common term of "contemporaneity" becomes incalculable. . . .

Certainly the migration of climatic belts and their dependent ecological zones during the Quaternary era gives similar successions of land floras and faunas in different regions. At first sight many might be regarded as coeval but detailed palaeontological, stratigraphic, and, today, radiometric studies have proved them otherwise. Here, however, we are dealing with a "*Feinstratigraphie*" not often encountered in the geological column. Few geologists would hold that fossil assemblages can indicate exactly the chronological identity of strata. Such assemblages certainly seem to point to an approximate time-range for the deposition of the containing rocks. The stratigrapher does not yet often think in terms of actual numbers of years when reckoning his geological dates and ages, though he speaks of "*Manticoceras* times," "*D. murchisoni* times," and so on.

The essential point to elude Huxley was that geological processes controlling the stratigraphical record are incomparably slower than biological migration and dispersal. If the migration or dispersal time for a species was at all comparable with that of the life of the species, then such organisms would be of no value in correlation. Many examples can be taken from the world around us to show without question that the life of a species is greater than the time required for migration (i.e., to cover its biological province).

Within the last 50 years a great deal has been learned about animal dispersal and migration and about the factors controlling them. Climatic factors are amongst the most important producing fairly well-defined belts on land and temperature zones in the sea. The control these (and physical barriers) exert is remarkably rigid and few organisms range at will from one belt to another. Climatic change will result in the wholesale migration of life, as witnessed during the Pleistocene glacial and interglacial phases.

In most, if not all, living communities there is a population pressure that encourages organisms to expand the area their species occupies and to migrate into new environments where possible. This is part of the regime tending to promote "the survival of the fittest" and to be part of the cause of organic evolution. It has clearly operated since living things first appeared on the earth.

The speed with which a newly introduced organism may spread throughout an environment in which it can survive is well exemplified by the occupation of Australia by the rabbit. Along the eastern littoral zone of North America the spread of the European periwinkle has similarly been very rapid. These are, of course, introductions made by man, but unexpected cases of such transfers from one area to another across oceanic areas or other barriers by natural—or at least non-human—means are known. The unexpected fauna of Madagascar has been attributed to the drifting of animals on rafts across deep water, and the Tertiary rodent populations were probably spread through the Caribbean area in the same manner. (Rodents still migrate in this way today on masses of sudd-like debris.)

The immigration of new genera into a region may leave us with a series of fossils for which we see locally no ancestors and can find no centre of radiation. These fossils have been called *cryptogenic*. For example, *Cryptolithus* appears first in Trentonian strata in North America: in Britain it is known from the base of the upper Llanvirn and presumably migrated from Europe to North America during the intervening time. The interval is a large one and may be reduced as further collections are made, but it is clear that the life span of this genus at an American locality may be very different from that in Britain. The Germans have employed the term *Teilzone* for the thickness of rock yielding a species at any one locality (its time range there being designated as a *Teilchron*): *Biozone* and *Biochron* correspond to the total range of rocks and time through which the index form (or forms) extends. Thus the American teilzone of *Cryptolithus* is different from that in Britain.

Unfortunately for Huxley's hypothesis, *Cryptolithus* does not seem to be a member of a whole succession of faunal assemblages migrating very slowly from the Old World to the New at about the same rate. Correlation between the European and the North American standard Ordovician sections can be maintained on very convincing general and particular grounds, including the graptolite sequence. Migration is recognized but it is an activity seen against a faunal background which provides ample data for correlation.

RELICT FAUNAS

The familiar indigenous faunas of Australia and New Zealand are modern examples of relict faunas. Their monotremes and marsupials were clearly isolated from the rest of the world at an early stage in mammalian evolution. Until historic times they were free from competition from placentals. A similar state of affairs existed late in parts of South America. How would we correlate the fossil remains of these faunas with those of land faunas elsewhere?

The marine fauna along the Australian coast south of 30° S. also shows peculiarities of the same kind. Here typically tropical genera extend into it more than into any other temperate zone, and some of them there give rise to their largest species; and here also are the last living occurrences of such forms as *Campanile, Gisortia,* and *Trigonia*. Are they to be regarded as "living fossils" giving the lie to the idea that such forms can be used in palaeontological correlation? In the Australian faunal province, separated by physical barriers from the rest of the world, the Mesozoic seems to hang on. It is, however, the presence of modern forms with relatively known Tertiary lineages or ancestry which establishes the obvious age of the fauna.

The Aralo-Caspian province includes an unusual fauna of marine mollusca associated with freshwater forms all of which have become adapted to a freshwater habitat since Miocene times. Lake Baikal retains, of all things, a seal now adapted to a non-marine habitat. The fauna here is another example of a relict association, cut off from the general run of evolutionary progress in the rest of the world (see Davies 1934), and a shift from one environment to another has gradually come about. It is the same, though perhaps more spectacular, with the protomollusc *Neopilina* and the coelacanth *Latimeria*.

The recent reports of graptolites from the Devonian of Bohemia attracts attention as a possible parallel. The Bohemian province appears as a distinctive realm from Ordovician until late in the Devonian but to what extent it was a refuge for "relict faunas" is a matter of some conjecture. Cretaceous conodonts are known only from Africa; Triassic conodonts in Europe, North America, and Spitsbergen seem to belong to a common fauna. The microfossils are the only members of this Cretaceous fauna which appear to be "relict," and here at least the stratigraphical range of conodonts is extended very considerably. Fortunately, the age of the deposit in which they occur can be substantiated without doubt by other fossils.

The question of "derived" fossils or faunas appears in the literature from time to time, and might be mentioned here. Certainly fossils derived from earlier horizons are well known in some geological formations and they are usually characterized by a worn appearance and are often fragmentary, phosphatized, or otherwise mineralized in a distinctive way. Some have been mistaken for relicts while some long-ranging species have been regarded as derived when first found beyond their previously accepted upper limit. The key in every case is careful examination and knowledge of the geological succession.

STATISTICAL ANALYSIS

Mathematical analysis of approximately corresponding faunas has been used in an effort to obtain precise correlation (Keen 1940). Here common species are counted from several levels in different stratal successions covering approximately the same time span and the faunas carrying the same percentages are then equated. It is a laborious method, full of hazards and pitfalls. Clearly it is useful in small areas but becomes unreliable over greater distances. Perhaps most familiar to students is the work of Lyell (1830–33) in attempting to divide the European marine Tertiary rocks on the basis of modern species present in the molluscan faunas. It was also used by Martin in his subdivision of the Tertiary of the Far East, a subdivision which proved to be remarkably well aligned with that of Europe (see Van der Vlerk 1959).

A. M. Davies (1934) quotes statistics from Clement Reid in a study of British Pliocene molluscan faunas, showing in successive horizons the decrease and disappearance of extinct and southern species and an increase in northern (coldwater) forms. Zeuner used the statistics to demonstrate the probable occurrence of two cold phases, during the deposition of the Newer Red Crag and of the Weyborne Crag.

Within recent years micropalaeontologists have been doing much the same sort of thing, especially in palynological studies. Much of their work has been with Pleistocene and Recent material and is of great value in demonstrating climatic change. Studies of evolution have been essential for the problem of subdivision of earlier rock formations, but for the Quaternary we need to replace the study of evolution by the study of climate.

CONCLUSION

All fossils have time ranges which enable them to be used to some degree as stratigraphical indices. Many have a short time-span and are excellent in this connection; others are geologically long-lived. Assuming that reasonable palaeontological identifications can be made, each and every fossil assemblage acquires relative geochronological significance, *provided* that its stratigraphical relationships are known. Guaranteed stratigraphical successions are absolutely essential before any palaeontological sequence can have meaning. The palaeontologist may have to grapple with problems of facies, migration, biological provinces, homeomorphs, and the like, but he has to rely primarily upon good stratigraphical field practice to give his faunal or floral record significance. His chronology is basically a relative scale and, despite both local and often as yet inexplicable vagaries, patently it works.

REFERENCES

ARKELL, W. J. (1933). The Jurassic System in Great Britain. Oxford: University Press.

ARKELL, W. J. *et al.* (1957). *In* Treatise on Palaeontology, pt. L, Mollusca 4, *by* R. C. Moore. Lawrence, Kansas: Geol. Soc. America.

BOND, G. (1950). Lower Carboniferous reefs in Northern England. J. Geol., *58*: 313–29.

BRINKMANN, R. (1929). Statistische-biostratigraphische Untersuchungen an Mitteljurassischen Ammoniten über Artbegriff und Stammesentwicklung. Abhandl. Ges. Wiss. Göttingen, Math-physik. Kl., Folge 3, no. 13: 1–250.

DAVIES, A. M. (1934). Tertiary Faunas. 2 vols. London: Murby.

HODSON, F. and LEWARNE, G. C. (1961). A mid-Carboniferous (Namurian) basin in parts of the counties of Limerick and Clare, Ireland. Quart. J. Geol. Soc. Lond., *117*: 307–34.

HUXLEY, T. H. (1862). The Anniversary Address. Quart. J. Geol. Soc. Lond., *18*: xlvi.

KEEN, A. M. (1940). The percentage method of stratigraphic dating. Proc. 6th Pacific Sci. Congr., 1939, ii: 659–63.

LOWENSTAM, H. (1950). Niagaran reefs of the Great Lakes area. J. Geol., *58*: 430–87.

LYELL, C. (1830–33). Principles of Geology, 1st ed., 3 vols. London: John Murray.

———— (1867). Principles of Geology, 10th ed. London: John Murray.

SIMPSON, G. G. (1944). Mode and Tempo in Evolution. New York: Columbia University Press.

TEICHART, C. (1958). Some biostratigraphical concepts. Bull. Geol. Soc. Am., *69*: 99–120.

VAN DER VLERK, I. M. (1959). Problems and principles of Tertiary and Quaternary stratigraphy (Fourteenth William Smith Lecture). Quart. J. Geol. Soc. Lond., *115*: 49–64.

VAUGHAN, A. (1905). The palaeontological sequence in the Carboniferous Limestone Series of the Bristol area. Quart. J. Geol. Soc. Lond., *61*: 181–305.

WHEELER, H. E. (1958). Time-stratigraphy. Bull. Am. Assoc. Retrol. Geol., *42*: 1047–1063.

ZEUNER, F. E. (1946). Dating the Past: An Introduction to Geochronology. London: Methuen & Co.

LIMITATIONS OF RADIOMETRIC DATING*

H. Baadsgaard, G. L. Cumming, R. E. Folinsbee, F.R.S.C., and J. D. Godfrey

When to the sessions of sweet silent thought
I summon up remembrance of things past,
I sigh the lack of many a thing I sought,
And with old woes new wail my dear times' waste.

WM. SHAKESPEARE

ABSTRACT

The four principal methods of radiometric dating are: uranium–thorium–lead, rubidium—strontium, potassium–argon, and carbon-14. The precision and reliability of the uranium lead and rubidium–strontium methods have been increased with the introduction of the concordia and isochron plots for data interpretation. The potassium argon method is the most versatile and has been applied successfully to minerals as young as a million years and as old as 4500 million, from a wide variety of plutonic, volcanic, tuffaceous, sedimentary, and meteoritic sources. The method gives reliable dates for certain minerals from unmetamorphosed terrains.

Specific new examples from the Precambrian Shield, the Cordillera, and the Western Canada basin have been used to illustrate geologic applications and limitations of the first three dating methods.

RADIOMETRIC DATING (Holmes 1962) is called on for the solution of problems in an increasing number of geologic settings. It is important that the geologist calling for such aid be aware of the limitations inherent in the methods applied, and that the radiometric specialist be aware of the degree of precision required by the geologist.

There have been a number of excellent review papers published in the past few years (Aldrich and Wetherill 1958; Tilton, 1960 a, b; Kulp 1961; Hamilton, Dodson, and Snelling 1962) on geochronology by radioactive decay, the comparison of isotopic dating methods, the geochronology of rock systems, and the application of physical and chemical methods to geochronology. It is not the intention of the writers to review the reviews, but rather, using specific geologic case histories from their own experience, to show some of the positive results obtained by the use of radiometric dating, and some of the time-consuming pitfalls into which they have tumbled.

Examples come in part from geologic problems posed by the writers' colleagues Drs. Burwash, Campbell, and Lerbekmo; others have been

*Work supported by the National Research Council of Canada, the Research Council of Alberta, the Geological Survey of Canada, and the University of Alberta Research Fund.

incorporated in theses by University of Alberta graduate students: Peterman (1962), Nascimbene (1963), and Shafiqullah (1963). The writers are grateful for permission to use unpublished material. The data have been presented in simplified rather than definitive form, as examples of the limitations of radiometric dating rather than as specific solutions to geologic problems.

Six criteria govern the successful application of radiometric dating to geologic problems (see Table I).

TABLE I

CRITERIA FOR RADIOMETRIC DATING

1. Decay constants must be known accurately.
2. Isotopic abundance of radioactive parent must be known and constant.
3. Sampling must be representative and adequate.
4. Analytical determinations must be of meaningful accuracy.
5. Only insignificant or correctable amounts of radiogenic daughter may be present.
6. Gain or loss of parent or daughter must be negligible.

DECAY CONSTANTS AND ABUNDANCES

It is only within the last decade that the two physical constants which govern radiometric chronometers, the decay constant and the atomic abundance of the radioactive parent, have been known with sufficient precision to make most age determinations (radiometric dates) meaningful. There is still not complete agreement between the various laboratories, particularly with regard to the atomic abundance and decay constants of K^{40} (Wetherill 1957) and the half-life of Rb^{87}, but the magnitude of disagreement at present results in differences in dates of less than 6 per cent.

Decay constants of radioactive Rb^{87} and K^{40} may be determined physically by counting methods (Glendenin 1961), or "geologically" by comparison of Rb–Sr and K–Ar results with the U–Pb dates of cogenetic uraninites (Wetherill et al. 1956). In general it has been noted that the use of physically determined decay constants tends to give lower radiometric dates by K–Ar and Rb–Sr methods than those determined using the "geological" constants. Undoubtedly there would be advantages to be gained by the adoption of a single set of decay constants and atomic abundances, in that interlaboratory comparisons would be facilitated (Wanless 1961; Afanas'yev et al. 1963).

It is unlikely that disagreement over constants (Table II) can be resolved completely within the next decade, and interim constants might best be adopted by international agreement. Somewhat the same problem besets the use of C^{14} as a geochronometer—recent physical measurements indicate that the widely adopted "Libby half-life" is in error by 3 per cent. Until the experimental measurements are concluded, however, published measurements based on the Libby constants are to be corrected by multiplying them by a 1.03 correction factor (Rubin 1962).

TABLE II

PRINCIPAL DECAY SERIES OF THE LONG-LIVED ISOTOPES

Parent	Daughter	Half-life, yr.	Isotopic abundance, at.%	Useful age range, m.y.
K^{40}	Ca^{40} β-decay	1.47×10^9	0.0118	
	Ar^{40} K-capture	1.19×10^{10}	or 0.0119	>0.1
Rb^{87}	Sr^{87}	4.7×10^{10}	27.85	>10
		or 5.0×10^{10}		
Th^{232}	Pb^{208}	1.39×10^{10}	100	
U^{235}	Pb^{207}	7.13×10^8	0.710	>10
U^{238}	Pb^{206}	4.51×10^9	99.285	

DATING DIABASE DYKES

In certain instances the samples collected in the course of routine reconnaissance geologic mapping will suffice for solution of a geologic problem. As an example one might consider the diabase dyke sets of the Canadian Shield. Gill and L'Esperance (1952) had concluded " . . . the diabase dykes may also have been intruded at many different times, though always at a late stage in an orogenic cycle and always in similar relations to the rocks and structures of a particular mountain belt. . . ." It seemed desirable to tests this hypothesis, using the K–Ar method, which has recently been applied with success to dolerites (McDougall 1963). In the course of field mapping in the Yellowknife geologic province the centres of large dykes had been sampled, also the chilled margin and the granite immediately adjacent to the dyke margin, to which the diabase had welded (Fig. 1).

Theoretically, the 2400-million-year-old biotite of the contact granite would have its radioactive clock "reset" at the time of dyke intrusion, through expulsion of radiogenic argon by heating. Biotite from the granite at the dyke contact returned a K–Ar date of 1010 million years, confirming the theory. The chilled diabasic margin of the dyke returned a K–Ar date of 970 million years, and the gabbroic dyke centre a date of 885 million years, suggesting that coarse whole rock gabbro is reasonably retentive of radiogenic argon and the finer dyke margin retains almost all its radiogenic argon.

Similar methods of dating were applied to 17 other diabase dyke samples, and from these emerged an age pattern in essential agreement with the Gill–L'Esperance hypothesis (Fig. 2). The dates will be of use in determining tectonic relations of major faults and the diabase dyke sets (Byers 1962). This intriguing picture is discussed in greater detail by Burwash et al. (1963). It is noteworthy that in the particular problem posed by the dyke sets a precision of 10 per cent is sufficient, and a difference of 100–200 million years in dates does not affect the geological interpretation significantly.

Although the use of whole rock samples has given apparently rewarding results in the case of diabase dykes, the writers do not give unqualified

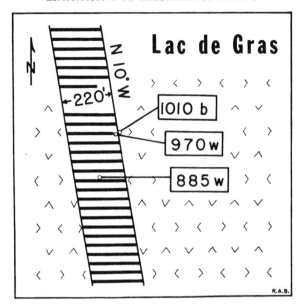

FIGURE 1. Diagrammatic sketch showing field relations of a diabase dyke intruding granite, for which the horn-felsitized biotite in the contact granite (1010 b), chilled diabase from the dyke contact (970 w), and coarse gabbro from the centre of the dyke (885 w) gave concordant K–Ar dates.

endorsement to this form of sampling. Feed forced indiscriminately into the maw of the time-machine can produce an undigestible mass of worthless data. The programme adapted by the Geological Survey of Canada (Lowdon 1960, 1961; Lowdon et al. 1963), in which carefully separated and analysed biotites have been dated, has produced a definitive, closely controlled picture of the development of the exposed portions of the Canadian Shield and Cordillera, whereas whole-rock samples from the same province produce a hopeless scatter in dates (Folinsbee 1955).

It was observed in Russia (Gerling et al. 1961) that some pyroxenes contain such large amounts of radiogenic argon that they might be considered to be of mantle origin. This phenomenon was confirmed by Hart and Dodd (1962), who concluded that pyroxenes may contain excess argon, and about the same time it came to the attention of the writers (Table III). It is clear that the matter of dating pyroxene-containing dykes must be approached with great caution but noteworthy that pyroxenes containing excess argon are potassium-deficient and do not contribute substantial amounts of radiogenic argon to the analysed whole-rock samples.

Thus criterion 6 (Table I)—that gain of the radiogenic daughter must be negligible—is important only in the case of pyroxenes of low potassium content and, as observed earlier, in potassium-deficient minerals such as

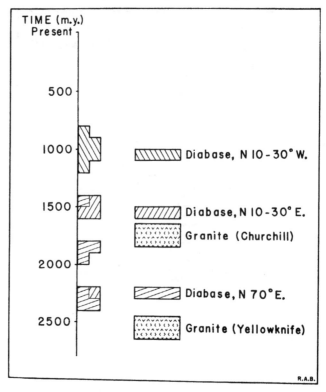

FIGURE 2. Periods of diabase dyke intrusion in the Yellowknife geologic province follow periods of orogeny and granite intrusion.

TABLE III

PYROXENES WITH EXCESS RADIOGENIC ARGON

	K_2O, %	Date, b.y.
Cobble, Tazin Conglomerate		
Hornblende	0.79	2.4
Pyroxene	0.04	6.4
East–West diabase, Noranda		
Pyroxene	0.015	5.6
Diabase dyke, Manitoba		
Whole rock	0.09	3.1
Pyroxene	0.04	3.3

beryl and cordierite whose atomic structure accommodates argon (Damon and Kulp 1958). Basic whole rocks do not appear to lose enough radiogenic daughter for the loss to be critical, though it is well known that the potash feldspars of granitic rocks lose argon, and that whole-rock K–Ar dates of granites and rhyolites are less than those for cogenetic micas (Folinsbee 1955).

DATING OF YOUNG SAMPLES

In another setting, differences in dates of a million years are of geologic interest. Radiometric dates for biotites from granites of the Cascade Range of the Cordillera (Baadsgaard, Folinsbee, and Lipson 1961) indicate the occurrence of an Alpine orogeny 18 million years ago. The writers were asked by Waters and his associates to date biotites from the Tatoosh pluton at Mt. Rainier, south of the Snoqualmie batholith (Fig. 3), to see if this pluton was Mio-Pliocene, and younger than the Snoqualmie batholith, as suggested by field relations (Waters 1961). K–Ar age determinations on the biotite substantiate the geologic hypothesis.

TABLE IV

ALPINE EPIZONAL GRANITES, CASCADE RANGE

Biotites	K_2O, %	Radiogenic argon, %	Date, m.y.
Hell's Gate	5.7	78	35
Chilliwack			
Wahleach	6.6	71	18
Cheam View	6.6	52	18
Snoqualmie	5.7	52	18
Tatoosh			
Goat Island	6.0	40	13
Mt. Rainier	7.6	35	15

In the determination of such young dates, criterion 5 (Table I)—that only insignificant or correctable amounts of radiogenic daughter may be present—becomes of great significance (Table IV). For a 35-million-year-old biotite date, 78 per cent of the argon recovered from the sample is radiogenic, but for the 13–15-million-year-old Tatoosh biotites only 35–40 per cent is radiogenic, and the problem of air argon correction becomes acute.

It is noteworthy that air argon contamination of sanidines is less than that of biotite, and that sanidines quantitatively retain their radiogenic daughter argon (Baadsgaard, Lipson, and Folinsbee 1961). There is some possibility that sanidine may be used for the determination of Pleistocene dates (Table V) with no greater air contamination than that for biotites 50 times older (Table V). The writers do not place great confidence in

TABLE V

LAACHER SEE VOLCANIC AREA, GERMANY
Pleistocene—Younger than Interstadial between Mindel 1 and Mindel 2

Sanidine	K_2O, %	Radiogenic argon, %	Date, yr.
Evernden et al. (1957)	13.7	46	370,000
Baadsgaard et al. (Alberta)	12.0	46	750,000

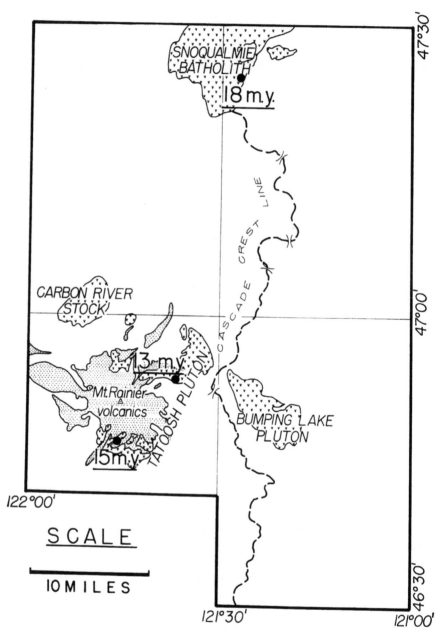

FIGURE 3. K–Ar dates confirm the geologic hypothesis that the Tatoosh pluton is somewhat younger than, but related to, the Mio-Pliocene Snoqualmie batholith (Waters 1961).

their single 750,000-year date obtained for sanidine from the Laacher See volcanics, and it is quoted only as an indication that Pleistocene K–Ar dates may be made with sufficient precision to be interesting.

RADIOMETRIC DATES FOR SAMPLES WITH BIOSTRATIGRAPHIC CONTROL

A different order of precision is required for meaningful dates on samples with good biostratigraphic control, and many of the enumerated limitations of radiometric dating (Table I) become critical. For a number of years the

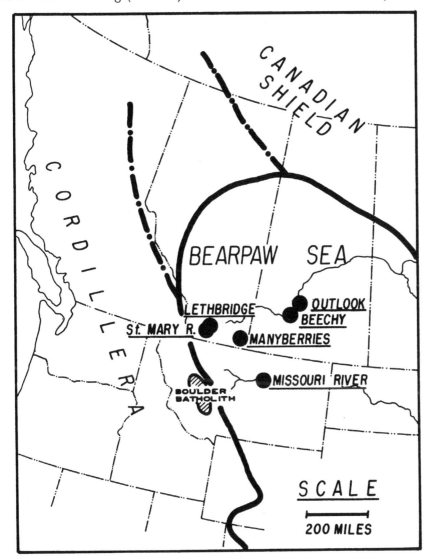

FIGURE 4. Datable bentonites occur in the basal Bearpaw marine shales over a wide area of the Great Plains region (Nascimbene 1963).

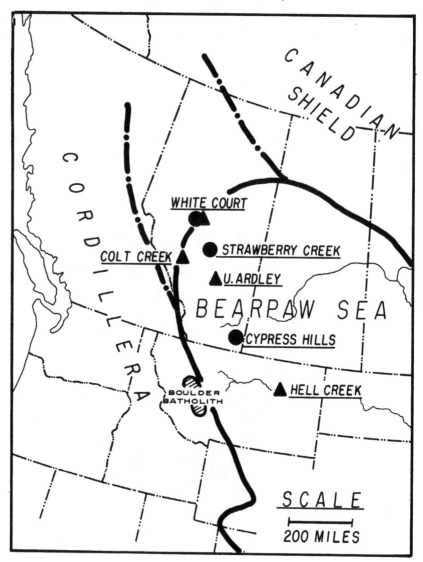

FIGURE 5. Kneehills tuff samples (Beginning Lance; circles) and Tertiary boundary coals (triangles) yield dates for the closing years of the Mesozoic era.

writers have been engaged in time-scale determinations—the assignment of precise radiometric dates to datable horizons where the biostratigraphic assignment is above reproach (Folinsbee *et al.* 1960, 1962). Certain key dates—the Mesozoic–Cenozoic boundary, for example (Watson 1956)— are particularly intriguing. In western Canada the waning years of the Mesozoic era were marked by vulcanism and the fall of bentonitic ash which became intercalated with marine shales (containing ammonites as the zone fossils) and continental sandstones (containing dinosaurs) (Figs. 4 and 5).

In determinations prior to 1962 the writers dated sanidine and biotite from the marine basal Bearpaw (Campanian) bentonites at Lethbridge as 75 million years old, with a standard deviation on nine runs of 3 million years. Subsequently the Bearpaw bentonites were sampled over a wider area geographically, and some of the samples lie slightly above the Lethbridge horizon stratigraphically. The newer dates are appreciably younger and have a smaller standard deviation (Table VI). With the present analytical precision in the writers' laboratory this average date for the basal Bearpaw of 72 million years, with a standard deviation of 3.3 million years, is a measure of the reproducibility that can be attained without further refinement of the sampling and analytical methods.

TABLE VI
BASAL BEARPAW BENTONITES
(Campanian)

	Biotite dates, m.y.	Sanidine dates, m.y.	Average dates, m.y.
Runs prior to 1962			
Lethbridge	75 68 77 74 74 74	76 77 77	75±3
Runs after 1962			
St. Mary River	71 71	71 68	
Manyberries	70 73 68 68		70±1.7
Outlook–Beechy	70 69		
Missouri River	72	69	

Average radiometric date = 72 m.y. ±3.3 s.d.

Slightly higher precision has been achieved in a determination of the date for the beginning of Lance (Maestrichtian) time—65.9+1.4 million years (Table VII). A date for the end of the Mesozoic has been arrived at from dating studies of samples from four localities (Table VIII), and a figure of 62–63 million years seems to be the best the writers have been able to obtain.

The writers have examined various possible causes of this lack of precision in dates. It does not seem to be a grain size effect (Table IX), and it is

TABLE VII
KNEEHILLS TUFF BENTONITES
(Lance or Maestrichtian)

	Biotite dates, m.y.	Sanidine dates, m.y.
Whitecourt	64.6	
Strawberry Creek	68.0 65.7 66.8 65.4	67.7 64.0 66.0 66.2
Cypress Hills		64.9

Average radiometric date = 65.9 m.y. ±1.4 s.d.

TABLE VIII
TERTIARY BOUNDARY COAL BENTONITES

	Biotite dates, m.y.			Sanidine dates, m.y.		Average dates, m.y.
Colt Creek (Paleocene)	59.6 59.8 58.7 58.7 60.0 59.3 61.5 61.1					59.8±1.0 s.d.
U. Ardley (Uppermost Cret.)	(52.9 51.5 49.2)			63.0		63.0
Whitecourt (Uppermost Cret.)	62.8			64.4		63.6
Hell Creek (Basal Ft Union/Paleocene)	60.7			64.4 61.1		62.1

TABLE IX
GRAIN SIZE EFFECT
Biotites from Paleocene Bentonites

Size fractions, mesh size	Dates, m.y.
60–120	59.6 60.0
120–170	59.8 59.3
170–270	58.7 61.1 61.5
270–325	59.9

possible to eliminate samples for which the weathering effect (Kulp and Bassett 1961) is noteworthy (Table X), by inspection of the potassium contents of the dated minerals. It is likely that the spread in dates represents the present state of the art, and is a measure of the combined effect of small errors in sampling (2, Table I) and analytical determinations (3, Table I). Nevertheless the determinations are becoming sufficiently precise to be of considerable interest to palaeontologists and stratigraphers.

TABLE X
WEATHERING EFFECT
Biotites from Paleocene Bentonites

	K_2O, %	Dates, m.y.		
Fresh	5.23	59.8	59.6	58.7
Fairly fresh	4.18	60.0	59.3	61.5
Slightly weathered	2.87	61.1	59.9	
Moderately weathered	1.99	58.7		
Strongly weathered	1.25	55.8		

Turning more towards analytical problems, one finds that points 4 and 5 (Table I) are related. The amount of daughter contaminant present in a mineral or rock phase at the time of formation of the phase must be small compared with the amount of radiogenic daughter, or the precision of a determined date will be strongly affected. Assuming that one can determine the amount of daughter contaminant to +1 per cent, the effect of this "error-in-correction" on the error in the amount of radiogenic daughter may be considered, using the Rb–Sr "chronometer" as an example (Table XI).

TABLE XI

EFFECT OF THE ERROR IN Sr^N DAUGHTER-CONTAMINANT DETERMINATION
ON THE PRECISION OF THE Sr^{87*} (RADIOGENIC) DETERMINATION FOR Rb–Sr DATING.
(For a sample containing 1 p.p.m. Sr^{87N} and 250 p.p.m. Rb^{87})

Date, m.y.	Sr^{87*}, p.p.m.	$Sr^{87*}/$(Total Sr^{87})$\times 100$, %	Percentage error in Sr^{87*} for a 1 per cent error in Sr^{87N} (± 0.01 p.p.m.)
6480	25	96	0.04
676	2.5	71	0.4
68	0.25	20	4
6.8	0.025	2.4	40

The phase chosen is a mica containing ~ 14 p.p.m. Sr^N (N = Normal) or 1 p.p.m. Sr^{87N} and ~ 900 p.p.m. Rb or 260 p.p.m. Rb^{87}. These are reasonable amounts for a granitic biotite. For an old sample of this mica it can be seen that the relatively low percentage of contaminant Sr^{87N} need not be known accurately in order to give an analytically reliable date. However, as the age (and thus the amount of radiogenic Sr^{87*}) decreases, the effect of a 1 per cent error in determining the contaminant Sr^{87N} becomes serious. Young mica samples must have very low Sr^{87N} contaminant if a reasonable date is to be obtained. For other minerals the situation may be even more critical. Feldspars often contain ten times more contaminant strontia than micas do, and usually contain less rubidium. If the Sr^N content is ten times larger than in the mica example, and the Rb content one-tenth as large, a feldspar 680 million years old would yield a date as much as 40 per cent in error, if 1 per cent error were made in the contaminant Sr^{87N} determination. The relative proportion of Rb to Sr and the age of the mineral are thus important criteria in selecting suitable samples for Rb–Sr dating.

Similar effects are present for the K–Ar and U–Pb chronometers, where old samples contain relatively small amounts of daughter contaminant, and the dates on progressively younger samples are less precise because of errors in daughter contaminant determination.

Another important factor in regard to daughter contamination is the isotopic composition of the contaminant element. In correcting for the daughter contaminant one most often assumes a definite isotopic composition for the contaminant element. Once the amount of a non-radiogenic isotope has been measured, the amount of daughter contaminant may be calculated by a simple ratio comparison. For K–Ar dating one uses the present isotopic composition of air argon in this computation. In old samples where one might expect large deviations in the composition of the primordial atmosphere (assumed contaminant), the atmospheric corrections are usually negligible. The assumption of atmospheric composition for the contaminant argon seems generally valid, judging from Rb–Sr dates on cogenetic samples. For Rb–Sr dating the composition of mixed contaminant strontia varies only slightly with geologic time, except in unusual cases.

Contaminant lead in uranium-bearing minerals may be "normal" lead or anomalous lead, and presents a problem if the composition of the con- taminant lead must be specified accurately. The amount of daughter contaminant varies for different minerals according to whether or not the mineral is a good geochemical "host" to the contaminant daughter element.

TABLE XII

MINERALS DATABLE TO ≤5 PER CENT PRECISION

Mineral	Minimum K–Ar date (approx.), m.y.	Minimum Rb–Sr date (approx.), m.y.	Mineral	Minimum U–Pb date (approx.), m.y.
Micas	1–2	80–100	Uraninite	10–20
Sanidine	0.1–0.2	300–400	Monazite	100–200
Hornblende	5–10	1000–1500	Zircon	300–400
Whole rock	5–10	500–1000	Allanite	1000–1500

The gain or loss of parent or daughter from the time of emplacement or deposition—criterion 6 (Table I)—is shown by unlike or discordant dates for different radiometric chronometers from the same rock sample. This sixth criterion is widely invalidated, yet (paradoxically) some of the most useful information of radiometric dating is obtained when different radio- metric chronometers yield discordant dates. In minerals containing a geo- chemically accommodated parent, the daughter element is not always an acceptable geochemical "guest." Because of this, and since discordant dates generally err in being apparently too young, the loss of daughter (or an intermediate in the case of the uranium series) has been used most widely to explain discordancy in radiometric dating. Metamorphism subsequent to emplacement or deposition of the rock is believed to be the major agency in daughter loss. Dates between the times of emplacement and metamor- phism may be obtained in the case of partial or low-grade metamorphism or low-temperature diffusion loss of the daughter, while recrystallization of the mineral may effectively "reset" the chronometer. This "smearing" of the original date constitutes one of the major problems in the interpretation of radiometric dates.

As an example of partial daughter loss, U–Pb data for a number of uranium-bearing minerals from a small area of the Precambrian shield in northeastern Alberta are shown on a "concordia" diagram in Figure 6. The curved line in the plot of Pb^{207}/U^{235} versus Pb^{206}/U^{238} is the locus of points which give the same dates for the two systems and is called the concordia. Lead loss or uranium gain yields points which plot below the concordia, while lead gain or uranium loss gives points which plot above the concordia. Wetherill (1956) has shown that variable lead loss caused by metamorphism at some time (t_2) after the initial crystallization of the mineral (t_1) should give values of Pb^{207}/U^{235} and Pb^{206}/U^{238} for various samples which lie on the line joining the two points on the concordia at t_1

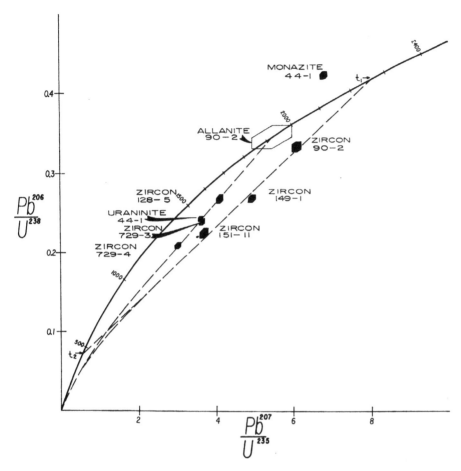

FIGURE 6. Parent-daughter ratios for U-bearing minerals from the Andrew Lake area, northeastern Alberta, compared with two curves calculated for loss of lead by continuous diffusion for 1920 m.y. and 2270 m.y.

and t_2. To account for deviations from an "episodic" lead loss model (as, for example, the present case, where no evidence for a suitable secondary metamorphic event can be found), Tilton (1960 a, b) assumed continuing diffusion of lead from the minerals and calculated the theoretical relationships. The dashed lines curving to the origin in Figure 6 are calculated Pb-diffusion-loss lines, and fit the given data quite well. The intersection of the diffusion-loss plot with the concordia gives the time of original crystallization. It should be mentioned that other data from other areas have shown an episodic loss pattern to which a diffusion pattern could not be applied.

Dates can be checked by employing a number of chronometers simultaneously, providing one knows to what extent each chronometer deviates from criterion 6 (Table I). The individual Pb-diffusion-loss dates, together with other radiometric data on other coexisting minerals in the same rocks, are given in Table XIII. From Table XIII it is seen that the mica K–Ar dates are uniform at ~1800 million years, the U–Pb dates on zircons give a double grouping, the U–Pb dates on secondary minerals fall in the younger zircon group, and the hornblende K–Ar and K–feldspar Rb–Sr dates lie between the two zircon groups. Pb–model ages on the Pb in the feldspars tend to follow the Rb–Sr dates as one might expect, while the galena date is grouped with the data for the secondary minerals. As regards changes in original dates, biotite is the most responsive mineral to metamorphism, hornblende and K–feldspar much less so, while the zircon chronometer is very difficult to reset completely (though Pb is apparently easily lost). A rough idea of the sequence of events is obtained when all the data are compared.

TABLE XIII

MINERAL DATES FROM A SMALL AREA IN THE PRECAMBRIAN SHIELD OF NORTHEASTERN ALBERTA

Rock	Dates, millions of years			
	U–Pb diffusion age	K–Ar	Rb–Sr	Russell–Stanton–Farquhar (1960) 207/206 model age
Basic plug	Zircon 1930	Biot. 1800 Hbl. 1840	—	—
Massive granite intrusive	Zircon 1960	Biot. 1790	—	—
"Older" granite	Zircon 1900	Biot. 1800 Hbl. 1910	K-feld. 1750	—
Granite boss	Zircon 2170	Boit. 1800	K-feld. 1700	—
Porphyroblastic granite–gneiss	Zircon 2230	Biot. 1800 Ser. 1740	K-feld. 2100	K-feld. 2160
Granite–gneiss basement	Zircon 2300	Biot. 1810 Hbl. 1930	K-feld. 1950 Whole rock 1920	K-feld. 2020
Secondary mineralization in metasediments	Uraninite 1900 Monazite 1920 Allanite 1920	—	—	Galena 1890

To get around the "smearing" of Rb–Sr dates, Compston and Jeffery (1961) proposed using a sample of the whole rock rather than a mineral separate. If one assumed short-distance chemical rearrangement within the rock sample during metamorphism, the whole-rock Rb–Sr date, being unchanged by internal variations, should indicate the time of emplacement of the rock. To make the data more reliable and delineate simultaneity of emplacement of the various phases of, in this instance, a basement gneiss, multiple samples with a range of Rb/Sr ratios are analysed and plotted as shown in Figure 7.

In Figure 7, Sr^{87}/Sr^{86} is plotted versus Rb^{87}/Sr^{86} to give a line whose slope is a function of Sr^{87}/Rb^{87}, or age, for contemporaneous samples. In the case of these samples from the basement gneiss complex, material of

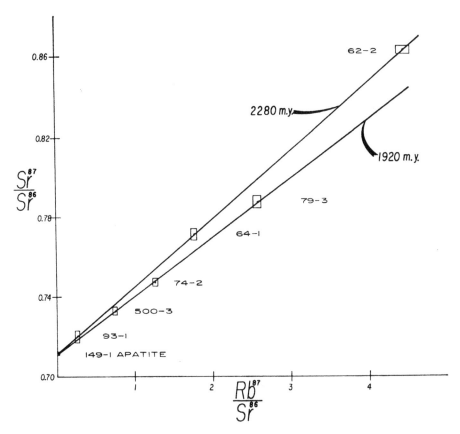

FIGURE 7. Plot of Sr^{87}/Sr^{86} ratios versus Rb^{87}/Sr^{86} ratios for six whole-rock samples and one apatite from the basement gneiss complex of the Andrew Lake area in northeastern Alberta.

two generations clearly shows up with dates very close to those given by the zircons on the U–Pb concordia plot (Fig. 6).

Tentative conclusions may be drawn from the data:

1. Two major magmatic or orogenic events took place ~1900 million years and ~2300 million years ago, and directly (magmatically) or indirectly (by sedimentation) contributed material to the area dated.

2. A wave of secondary mineralization was associated with the 1900-million-year event.

3. A last possible event is indicated by the ~1800-million-year biotite

date. This may reflect unroofing and dynamic metamorphism rather than thermal metamorphism.

GENERALIZED SUMMARY OF LIMITATIONS OF ISOTOPIC DATING

With the present precision, the absolute magnitude of the analytical error in a Precambrian date is often of the order of 50–100 million years. Events which may take place consecutively within this period of time cannot yet be resolved. Although the absolute magnitude of the analytical precision decreases with decreasing age of sample, errors due to daughter contamination eventually rise to a level equivalent to the gain in "resolution" between two geologic (clock "resetting" or "updating") events. In addition, the relatively small amount of radiogenic daughter lead or strontium makes the applicability of the U–Pb and Rb–Sr chronometers to Mesozoic rocks very limited.

Disagreement or uncertainty in the decay constants and isotopic abundances is no longer a major analytical limitation in radiometric dating.

Details of the process of rearrangement of isotopes during metamorphism are still poorly understood. For example, if low-temperature (100° C) diffusion of Pb gives discordant U–Pb dates, one should not expect to find the number of concordant U–Pb dates that do occur. The wide variety of possible chemical rearrangements in rock systems permits variable interpretation of discordancy patterns. The relative effects of dynamic versus thermal metamorphism on various parent–daughter systems are as yet undefined.

Since the duration of most geologic processes precludes the study of isotope migration under comparable conditions in the laboratory, one must use field interpretations and geologic evidence to explain many aspects of discordancy. Some of this ancillary evidence gives rise to circular reasoning, giving these interpretations of discordancy an unsound basis. In addition, inadequate geologic evidence may compound the difficulties.

The collecting of datable rocks carefully selected on the basis of the best field evidence is necessary. Mineral separations must then be carefully made with a view towards minimizing contamination, obtaining as pure a phase as is possible, selecting the most suitable mineral combinations for dating, and maintaining the parent–daughter ratio during the separation. Sampling for dating must take into account the problems previously touched upon, or the full (and necessary) mutual support of the radiometric and geologic evidence will not be attained.

Interpretations of radiometric data obtained from a few samples using one method should be generalized and cautious at the present time, especially in areas where subsequent metamorphism is seen or suspected. The use of as many methods of isotopic dating as feasible gives the most suitable data for interpretation. For geologically young samples only K–Ar dating can be applied with the requisite precision.

REFERENCES

AFANAS'YEV, G. D., BORISEVICH, I V., SHANIN, L. L., and SHEINA, I. P. (1963). Cases of unbalanced K–Ar relationships in biotites in connection with the construction of a geochronologic scale in an absolute time-table. Izv. Akad. Nauk SSSR Ser. Geol., 1: 19–45.

ALDRICH, L. T. and WETHERILL, G. W. (1958). Geochronology by radioactive decay. Ann. Rev. Nucl. Sci., 8: 257–98.

BAADSGAARD, H., FOLINSBEE, R. E., and LIPSON, J. (1961). Potassium–argon dates of biotites from Cordilleran granites. Bull. Geol. Soc. Am., 72: 689–702.

BAADSGAARD, H., LIPSON, J., and FOLINSBEE, R. E. (1961). The leakage of radiogenic argon from sanidine. Geochim. Cosmochim. Acta, 25: 147–57.

BURWASH, R. A., BAADSGAARD, H., CAMPBELL, F. A., CUMMING, G. L., and FOLINSBEE, R. E. (1963). Potassium–argon dates of diabase dyke systems, District of Mackenzie, N.W.T. Trans. Can. Inst. Min. Met., 66: 303–7.

BYERS, A. R. (1962). Major faults in western part of Canadian Shield with special reference to Saskatchewan. In The Tectonics of the Canadian Shield, edited by J. S. Stevenson (Roy. Soc. Can. Spec. Pub. no. 4), pp. 40–59. Toronto: University of Toronto Press.

COMPSTON, W. and JEFFERY, P. M. (1961). Metamorphic chronology by the rubidium–strontium method. Ann. N. Y. Acad. Sci., 91 (2): 185–91.

DAMON, P. E. and KULP, J. L. (1958). Excess helium and argon in beryl and other minerals. Am. Mineralogist, 43: 433–59.

EVERNDEN, J. F., CURTIS, G. H., and KISTLER, R. (1957). Potassium–argon dating of Pleistocene volcanics. Quaternaria, 4: 13–17.

FOLINSBEE, R. E. (1955). Archean monazite in beach concentrates, Yellowknife geologic province, Northwest Territories, Canada. Trans. Roy. Soc. Can. Ser. 3, Sec. IV, 49: 7–24.

FOLINSBEE, R. E., BAADSGAARD, H., and LIPSON, J. (1960). Potassium–argon time-scale. Intern. Geol. Congress, XXI Session, Norden, pt. III, pp. 7–17.

FOLINSBEE, R. E., BAADSGAARD, H., and CUMMING, G. L. (1962). Dating of volcanic ash beds (bentonites) by the K Ar method. Nuclear Geophysics, Nuclear Science Series Rept. 38, Nat. Acad. Sci.–Nat. Res. Council, Pub. 1075: 70–82.

GERLING, E. K., SHUKOLYUKOV, Yu. A., KOLTSOVA, T. V., and MATVEYEVA, I. I. (1961). Age of the basic rocks of the Monchegorsky pluton. Rept. on the 10th Ann. Meeting of the Commission for the Determination of the Absolute Age of Geologic Formations, as prepared by H. Faul for the National Science Foundation, pp. 7–8.

GILL, J. E. and L'ESPERANCE, R. (1952). Diabase dykes in the Canadian Shield. Trans. Roy. Soc. Can., Ser. 3, Sec. IV, 46: 25–36.

GLENDENIN, L. E. (1961). Present status of the decay constants. Ann. N. Y. Acad. Sci., 91 (2): 166–80.

HAMILTON, E. I., DODSON, M. H., and SNELLING, N. J. (1962). The application of physical and chemical methods of geochronology. Intern. J. Appl. Radiation Isotopes, 13: 587–610.

HART, S. R. and DODD, R. T., JR. (1962). Excess radiogenic argon in pyroxenes. J. Geophys. Res., 67: 2998–9.

HOLMES, A. (1962). 'Absolute' Age: A Meaningless Term. Nature, 196: 1238.

KULP, J. L. (Editor) (1961). Geochronology of rock systems. Ann. N. Y. Acad. Sci., 91 (2): 159–594.

KULP, J. L. and BASSETT, W. H. (1961). The base-exchange effects on potassium–argon and rubidium–strontium isotopic ages. Ann. N. Y. Acad. Sci., 91 (2): 225–6.

LOWDON, J. A. (1960, 1961). Age determinations by the Geological Survey of Canada, Reports 1 & 2, Isotopic Ages. Geol. Surv. Can., Papers 60-17, 61-17.

LOWDON, J. A., STOCKWELL, C. H., TIPPER, H. W., and WANLESS, R. K. (1963). Age determinations and geological studies. Geol. Surv. Can. Paper 62-17.

MCDOUGALL, I. (1963). Potassium argon age measurements on dolerites from Antarctica and South Africa. J. Geophys. Res., 68: 1535–45.

NASCIMBENE, G. (1963). Bentonites and the geochronology of the Bearpaw sea. Unpubl. M.Sc. Thesis, University of Alberta.

PETERMAN, Z. E. (1962). Precambrian basement of Saskatchewan and Manitoba. Unpubl. Ph.D. Thesis, University of Alberta.

RUBIN, M. (1962). The Fifth International Conference on Radiocarbon Dating. GeoTimes, 7 (4): 32.

RUSSELL, R. D. and FARQUHAR, R. M. (1960). Lead Isotopes in Geology, pp. 52–58. New York: Interscience.

SHAFIQULLAH, M. (1963). Geochronology of Cretaceous Tertiary boundary, Alberta, Canada. Unpubl. M.Sc. Thesis, University of Alberta.

TILTON, G. R. (1960a). Volume diffusion as a mechanism for discordant lead ages. J. Geophys. Res., 65: 2933–45.

——— (1960b). Comparison of isotopic dating methods. Summer Course on Nuclear Geology, Varenna, Comitato Nazionale per L'Energia Nucleare, Laboratorio di Geologia Nucleare, Pisa, pp. 331–45.

WANLESS, R. K. (1961). Geochronology of rock systems. Ann. N. Y. Acad. Sci. 91 (2): 361.

WATERS, A. C. (1961). Keechelus problem, Cascade Mountains, Washington. Northwest Sci., 35 (2): 39–57.

WATSON, D. M. S. (1956). The two great breaks in the history of life. Quart. J. Geol. Sci. London, 112: 435–44.

WETHERILL, G. W. (1956). Discordant uranium–lead ages, I. Trans. Am. Geophys. Union, 37: 320–6.

WETHERILL, G. W. (1957). Radioactivity of potassium and geologic time. Science, 126: 545–9.

WETHERILL, G. W., WASSERBURG, G. J., ALDRICH, L. T., TILTON, G. R., and HAYDEN, R. J. (1956). Decay constants of K^{40} as determined by the radiogenic argon content of potassium minerals. Phys. Rev., 103: 987–9.

PALAEOMAGNETISM AS A MEANS OF DATING GEOLOGICAL EVENTS

L. W. Morley and A. Larochelle

ABSTRACT

In principle there are three distinct aspects of the palaeomagnetic method that may be considered in attempting to date geological events: the palaeomagnetic evidence for polar wandering allows geological dating on a coarse time-scale; the numerous polarity reversals observed in rocks suggest that it may become possible to date their magnetization with an accuracy of the order of one million years; the scatter in magnetization directions observed in penecontemporaneous rocks resulting from secular field variations might allow, in future, the dating of geological events to about the nearest thousand years.

In practice many difficulties are encountered in attempting to follow any of these approaches for dating rocks or geological events. The inaccuracy and incompleteness of the existing polar wandering curves are pointed out as the main shortcoming in the first case. A brief analysis of the existing North American polar wandering curves reveals that considerable work has yet to be done before the potentialities of this method can be estimated. Similarly, a complete and accurate survey of the earth's field reversals in the past will be required before rocks can be dated with a relatively high precision on the basis of their polarity. A method of estimating the time span between successive reversals is suggested. The possibilities of using the scatter in the magnetization direction of penecontemporaneous rocks, although attractive, do not seem to be very hopeful for rocks older than several thousand years.

THE SCIENCE of palaeomagnetism evolved from studies of rock magnetism when two facts became evident. The first is that certain igneous and sedimentary rocks are found to be permanently magnetized in the direction of the ambient earth's magnetic field at the time of their formation; the second is that this direction is commonly different from that of the earth's field now, and that in general the older the rock is, the greater is this difference. Further investigations revealed that accompanying this so-called polar wandering was another characteristic of the earth's field: it appears to have reversed its polarity several times in the geological past. Furthermore, from compass measurements over the past 500 years or so, the magnetic north pole appears to be circling about the geographic pole with a radius of about 25 degrees, giving rise to the phenomenon known as secular variation. Thus the earth's magnetic field seems to have varied its direction throughout geological history in three distinct ways: first by secular variation, secondly by reversal of polarity, believed to have a period of the order of about one million years, and thirdly by polar wandering at an estimated average rate of 0.3 degree per million years.

The Use of the Palaeomagnetic Polar Wandering Curve

For the geochronologist the polar wandering aspect of the earth's field seems to hold the most promise. Assuming that the earth's field in geological history approximated, as it does at present, that of a dipole at the centre of the earth, it is possible from the measurement of the palaeomagnetic inclination and declination, in a suitable rock sample at any known latitude and longtitude, to calculate the location of the pole at the time the rock was formed. If, in this way, the palaeomagnetic poles are calculated for undisturbed rocks of various known ages, it is possible to plot what is termed a "polar wandering" curve. After such a curve has been established reliably, it should then be possible to determine the ages of other rocks by locating their palaeomagnetic poles on the calibrated polar wandering curve. In essence this is the method of age determination by the palaeomagnetic pole wandering technique. It should be pointed out, however, that in the process of constructing a polar wandering curve and later in determining ages of unknown rocks from this, it is necessary to use rocks from the same continent. This has been found necessary because polar wandering curves of rocks from different continents, although having the same general shape, are nevertheless displaced by great distances. It was the discovery of this fact that led to the revival of the continental drift theory.

Reliability of Poles and Polar Wandering Curves

The basis of comparison being the polar wandering curve, the method cannot be more reliable than the curve itself. For this reason, it is imperative to evaluate the validity of the existing polar wandering curves and to examine the possibility of improving them.

The reliability of an existing polar wandering curve can best be evaluated by comparison with what might be considered the ideal. Such a curve would be based on a very large number of points representing rocks suitable for palaeomagnetic purposes and of many ages. Furthermore, these ages should be known indisputably. Obviously, the hope of obtaining even a sufficient number of points from each geological period, especially the older, is probably Utopian.

In an article written in July 1960, Irving (1961) stated: "since the beginning of 1959 about seventeen papers have appeared containing 54 pole determinations, and in only 16 of these determinations is any direct evidence of stability presented." If this statement alone is enough to put in doubt the validity of the polar wandering curves now in existence, there are other reasons that may be added to confirm these doubts. Some of these reasons are mentioned in the next few paragraphs.

It is not sufficient that the rocks dealt with in palaeomagnetic studies be magnetically stable. The fundamental question is this: Does the direction of magnetization arrived at in the laboratory actually reflect the attitude of the ambient geomagnetic field at the time the rock was first polarized magnetically? For the answer to this question to be confidently in the affirmative, it must be shown that the rock is free from secondary foliation, or schistosity. Similarly, rocks containing both primary and secondary ferromagnetic minerals must be eliminated from palaeomagnetic studies for the obvious reason that part or all of their polarization has probably been acquired under the influence of a geomagnetic field whose orientation was completely different from that of the field prevailing at the time the rock was first consolidated. Secondary minerals are most commonly encountered in rocks that have been exposed to elevated temperatures and pressures, and, for this reason, metamorphic rocks must be rejected systematically from consideration.

Many pole positions may have erroneously been regarded as reliable because neither their magnetic stability nor their suitability in terms of their petrologic character can be challenged although their age or position in the biostratigraphic column has never been established reliably.

Proper choice of sampling sites and rock types is most important. It has been estimated, for example, that 82 per cent of all the North American pole positions were derived from sedimentary rocks—mostly from red beds. There is no denial that red beds are probably suitable for this type of work as far as their stability and fossil-bearing character are concerned, but their deposition and magnetization may not have been contemporaneous.

Many points on the existing polar wandering curves are probably inaccurate because they are based on inadequate or insufficient sampling. The use of an orienting device other than the ordinary magnetic needle compass may not be required for collecting samples of many rock types, but for strongly magnetic rocks, such as certain volcanic types, a serious error in orienting the sample can be introduced unless the latter is oriented directly with respect to a geographic meridian, by means of a solar compass for instance. It is also important in such rocks to take a sufficiently large number of samples to average out the effect of local anomalies. A large number of samples (not less than about 30) is also important in order to eliminate the effect of secular variation. In a formation which may have taken more than 500 years to cool, the effect of secular variation could conceivably account for a spread as much as 30 degrees in the individual determinations.

A number of points on the existing polar wandering curves are based on samples of unconsolidated deposits, such as varved clays. There seems to be little doubt that these deposits acquired their present magnetization at the time of deposition, but the alignment of magnetization in the direction of the ambient geomagnetic field is not so certain. Experiments carried out a few years ago by King (1955) have shown that in the process of settling of magnetic sand in a tank, the magnetic axes of the particles have a tendency to orient themselves along the meridian of the ambient field but, because of the geometric form of the grains, at a shallower inclination than that of the field. It is therefore clear that points on the polar wandering curve that are based on samples that have acquired their magnetization during a settling process may be misleading.

Many rocks intrusive into metamorphic terrain of complex structure are surprisingly fresh and seem to meet all the requirements in regard to magnetic stability. However, because of the absence of sedimentary rocks in the vicinity, their attitude with reference to the "palaeohorizontal plane" is not defined, and hence pole positions calculated from such rocks should be regarded with suspicion.

The scope of the present paper does not allow a systematic review of all the palaeomagnetic measurements used so far for building up the polar wandering curves for all the continents of the world. Accordingly, we shall restrict ourselves to a brief analysis of the North American polar wandering

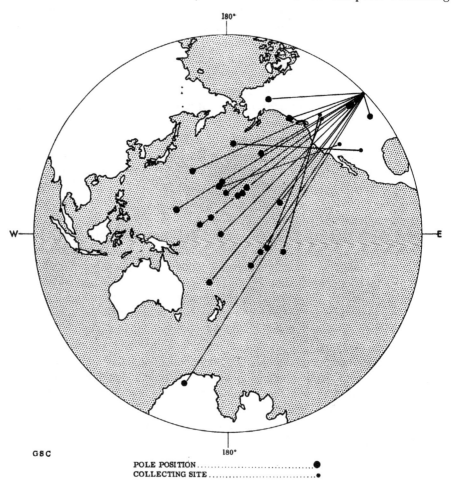

FIGURE 1.   Proterozoic pole positions as derived from North American rocks.

curves as a means of evaluating the potentialities of palaeomagnetism as a method of geochronology in North America.

In Figures 1 to 8, North American poles that have appeared in the literature, which was reviewed by Cox and Doell (1960), are plotted for geological periods or groups of periods together with the corresponding locations of the sampling sites.

Proterozoic (Fig. 1). A relatively large number of pole positions have been derived from Proterozoic rocks collected at widely separated localities in North America. It is doubtful whether a complete polar wandering curve for the Precambrian will ever be compiled because comparatively little of the bedrock is left in its original state. Also it is believed that the Precambrian lasted some three billion years so it is probable that the polar wandering curve crossed over itself several times at one or more points on the globe during that interval. Nevertheless, important uninterrupted

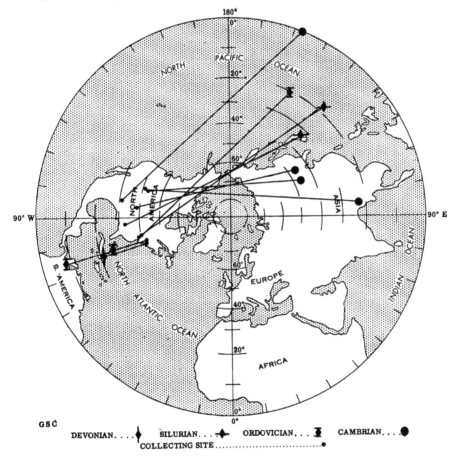

FIGURE 2.   Pre-Carboniferous Palaeozoic pole positions as derived from North American rocks.

sections of this period have been or could be studied from cratons overlain by relatively undisturbed and unmetamorphosed Proterozoic rocks. A better opportunity perhaps lies in the sampling of the ubiquitous diabase dykes many of which have apparently not been disturbed since their emplacement. Since present indications show, through the use of isotopic methods, that these dykes span the Proterozoic in time, it should become possible to trace the polar wandering curve for much of this geological era.

Pre-Carboniferous Palaeozoic (Fig. 2). The scatter displayed by the few available data for the Cambrian poles would be enough to discourage any effort to use palaeomagnetism as a means of dating early Palaeozoic rocks were it not for the questionable validity of some of these poles. Less scatter characterizes the group of published Ordovician poles, although it is to be noted that these poles are based exclusively on sedimentary rocks, some of which are strongly suspected of having acquired their magnetization long after deposition. Only two Silurian North American poles have been

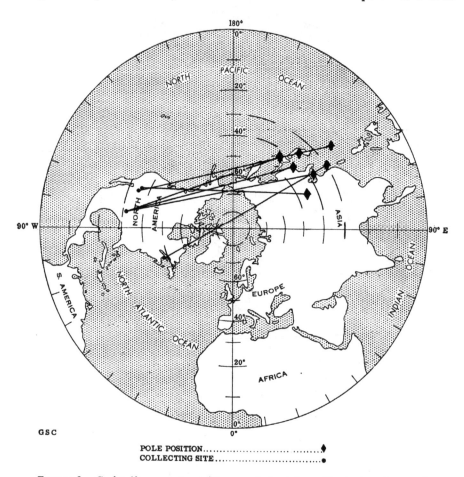

FIGURE 3. Carboniferous pole positions as derived from North American rocks.

published, and, despite their relatively good coincidence, the contemporaneity of their magnetization and deposition is in serious doubt. Finally, the only North American Devonian pole available in the literature is based on rocks for which the origin and age of the magnetization are uncertain. Thus the scarcity of results available for Early Palaeozoic North American rocks does not allow us to determine the age of the rocks on the basis of their palaeomagnetism. However, many well-dated formations of Early Palaeozoic age are known throughout Canada, and these could be sampled to redefine the polar wandering curve.

Carboniferous (Fig. 3). In contrast to the Early Palaeozoic poles, the Carboniferous poles derived from North American rocks are relatively closely grouped. It must be recognized, however, that these poles have all been derived from sedimentary rocks, and measurements on volcanic rocks are still needed to confirm their validity.

Permian (Fig. 4). Particular attention has been given to the palaeomagnetism of the Supai Formation whose deposition transgresses the Permo-

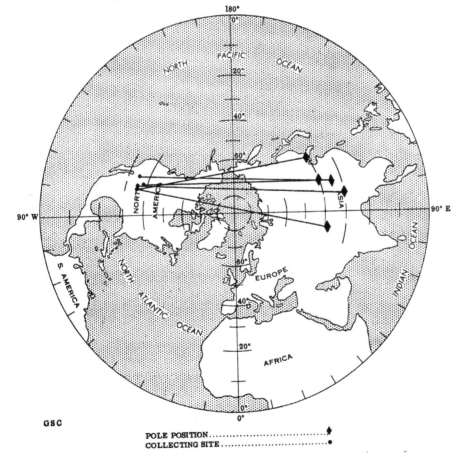

FIGURE 4.   Permian pole positions as derived from North American rocks.

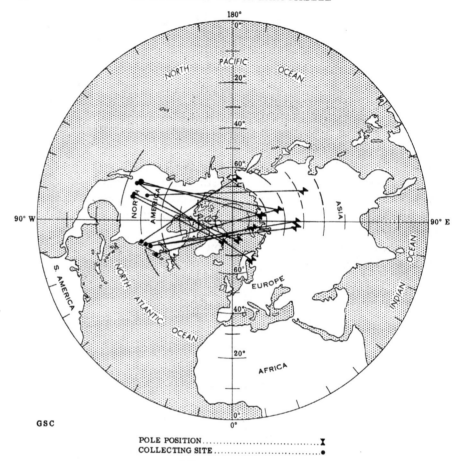

GSC

POLE POSITION..................................**X**
COLLECTING SITE.............................**●**

FIGURE 5. Triassic pole positions as derived from North American rocks.

Carboniferous boundary. Three groups of workers collected oriented samples at widely separated occurrences of this formation. The pole position that each group arrived at is very close to the mean pole derived from the totality of the samples collected, i.e. at 40° N. and 110° E. Three other Permian formations have been sampled at either one or two sites each. Although the results obtained are in perfect agreement with the indications of the Supai Formation, their validity would be greatly enhanced by more extensive sampling. From the data available at the present time, however, it does seem fairly certain that the north pole migrated from a location centring on Korea in Carboniferous times to one near Central China in Permian times, a distance of approximately 3000 miles. However, more data are required to confirm this section of the path.

Triassic (Fig. 5). Several North American Triassic formations have been sampled extensively at many sites, but the wide scatter of the resulting pole positions is disconcerting. They are spread between latitudes 50° N. and 80° N. around the present geographic pole. It seems that a complete

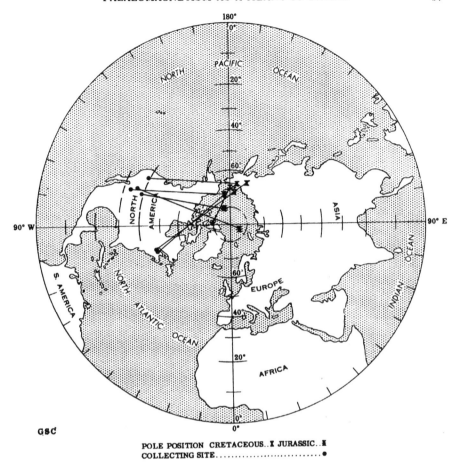

FIGURE 6.   Cretaceous and Jurassic pole positions as derived from North American rocks.

re-examination of the existing data is necessary as well as the addition of new data.

Jurassic and Cretaceous (Fig. 6). The Jurassic section of the North American polar wandering curve is very sketchy since it is based on the sampling of only two formations, one of which was sampled at a single site. A better representation is available for the Cretaceous period from the extensive sampling of four widely separated rock units comprising sedimentary, volcanic, and intrusive rocks of both polarities. The available Cretaceous poles are particularly well clustered around their mean.

Tertiary (Fig. 7). The number of poles calculated for different formations of Tertiary and Early Pleistocene age is probably sufficient to leave little doubt that the earth's dipole axis was inclined only slightly to the present geographic axis during this period. The validity of the data is enhanced by the fact that reversed polarites are a common feature in many of the formations of various types.

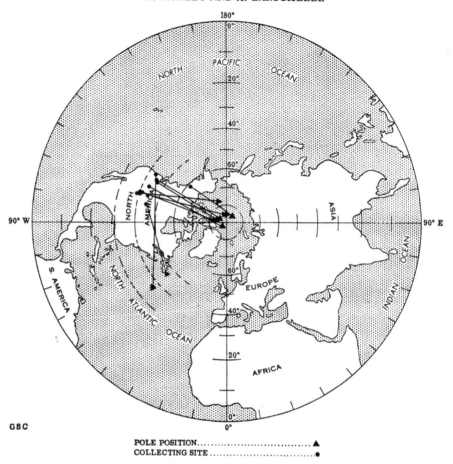

GSC

POLE POSITION.................................▲
COLLECTING SITE...............................●

FIGURE 7. Tertiary pole positions as derived from North American rocks.

Recent (Fig. 8). Apart from the observatory records of the last few centuries there are only two pole positions for this period which have been derived from North American rocks. As in other continents, the mean value of these poles coincides closely with the present geographic pole, and their magnetization directions follow the present polarity of the earth's field.

A more extensive review of the North American palaeomagnetic literature is hardly needed to show the necessity for more palaeomagnetic measurements throughout the geological column, as well as improvements, or at least standardization, in the methods of sampling, measuring, and testing, before the polar wandering concept of palaeomagnetism may be considered as a method of geochronology. The most pressing need for new data would appear to be for Proterozoic, pre-Carboniferous Palaeozoic, and Triassic times.

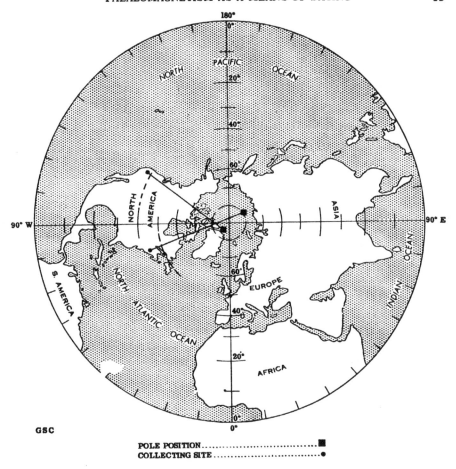

POLE POSITION.................................■
COLLECTING SITE.............................●

FIGURE 8.   Late Pleistocene and Recent pole positions as derived from North American rocks.

THE USE OF REVERSALS AS A MEANS OF GEOCHRONOLOGY

The fact that the earth's magnetic field has reversed its polarity, perhaps more than a hundred times since the Carboniferous (R. L. Wilson 1962), has been a difficult concept for many to accept. The evidence, however, is now overwhelming, and we do not intend to review it here. Unfortunately, so little is known about the times and exact durations of these reversals that they are of little use as geochronological markers. On the other hand, these palaeomagnetic reversals are found to be of great help in problems of correlation in non-fossiliferous rocks. Examples of this type of work are Einarsson's (1957) Magneto-Correlation in Icelandic lavas. Black (1963) found a large number of reversals in the Belt Series of southwestern Alberta that would be very useful for correlation purposes in these rocks, which are quite devoid of fossils useful for correlation.

The possibility of calibrating the time of these reversals throughout several periods of geological history is remote, especially since absolute age methods are at present not a great deal more definite than to the nearest geological period. However, Morley suggests* that a nearly unbroken record of these reversals may exist in the permanent magnetization of the rocks on the floor of the ocean basins. Mason and Raff (1961) have outlined a wide sequence of long north–south striking magnetic anomalies whose origin is puzzling. If one refers to recent papers of several authors (J. T. Wilson 1963; Vacquier 1962; Dietz 1961) in which the concept of mantle convection currents rising under ocean ridges, travelling horizontally under ocean floors, and sinking at ocean troughs is presented, it stands to reason that the rising rock, as it reaches the Curie point geotherm, will become permanently magnetized in the direction of the earth's field prevailing at the time. As this portion of rock moves upward and then horizontally to make room for the following rising material, and if, in the meantime, the earth's field has reversed and the same process is repeated many times throughout geological history, a linear magnetic anomaly pattern of the type observed would result. The present rate of travel of this mantle convection is believed to be of the order of a few centimeters per year. If this figure could be determined accurately for the various geological periods, it would then be possible, using the data from a magnetometer survey of the ocean basin floors, to reconstruct not only the history of reversals of the earth's field but also the direction and rate of the drift of continents. This time-reversal information could then be applied to problems of geochronology on the continents. Although such thinking is still very speculative, the study of palaeomagnetism and magnetometer surveys in the oceanic islands and ocean basins will no doubt prove or disprove it in the near future.

Geochronology by Palaeomagnetic Secular Variation

From observatory records it appears that the magnetic dip pole wanders within a circle of about 25 degrees radius centred at the geographic pole, so that it may be possible by this method to date material that exhibits stable magnetic remanence. Indeed the method has been used by Thellier and Thellier (1959) for dating ancient pottery made of baked clays. Attempts have also been made with less success to date recent varved clays by this method (Johnson, Murphy, and Torreson, 1948). The possibility of being able to combine such data with age determinations by the carbon-14 method is promising. It does seem quite clear, however, that there will be

---

*This hypothesis was originally presented at the Annual Meeting of the Royal Society of Canada in Quebec City on June 4, 1963. Since then, a letter by F. J. Vine and D. H. Matthews, Department of Geodesy and Geophysics, University of Cambridge, entitled "Magnetic Anomalies over Oceanic Ridges" has been published in *Nature* (No. 4897, Sept. 7, 1963, pp. 947–9), in which a similar hypothesis is presented. It appears that similar conclusions were arrived at independently.

little possibility of applying this technique to materials older than a few thousand years.

CONCLUSIONS

There is no doubt that palaeomagnetism will never supplant any other methods of geochronology in use today, but this is not the ambition of the palaeomagnetist. It is hoped, nevertheless, that some use will be made of a record still preserved in many rocks for filling gaps that cannot be filled by other methods or in supporting one or other of these methods in cases of disagreement.

REFERENCES

BLACK, R. F. (1963). Palaeomagnetism of part of the Purcell System in southwestern Alberta and southeastern British Columbia. Geol. Surv. Can., Bull. 84.

COX, A. and DOELL, R. R. (1960). Review of palaeomagnetism. Bull. Geol. Soc. Am., 71: 645–768.

DIETZ, R. S. (1961). Continent and ocean basin evolution by spreading of the sea floor. Nature, 190: 854–7.

EINARSSON, T. (1957). Magneto-geological mapping in Iceland with the use of a compass. Advan. Phys., 6: 232–9.

IRVING, E. (1961). Palaeomagnetic method: A discussion of a recent paper by A. E. M. Nairn. J. Geol., 69: 226–31.

JOHNSON, E. A., MURPHY, T., and TORRESON, O. W. (1948). Prehistory of the earth's magnetic field. Terr. Mag., 53: 349–72.

KING, R. F. (1955). The remanent magnetism of artificially deposited sediments. Monthly Notices Roy. Astron. Soc. (Geophys. Ser.), 7: 115–34.

MASON, R. G. and RAFF, A. D. (1961). Magnetic survey of the West Coast of North America 32° N Latitude to 42° N Latitude. Geol. Soc. Am. Bull., 72: 1259–70.

THELLIER, E. and THELLIER, O. (1959). Sur l'intensité du champ magnétique terrestre dans le passé historique et géologique. Ann. Géophys., 15 (3): 1–92.

VACQUIER, V. (1962). Magnetic evidence for horizontal displacement in the floor of the Pacific Ocean. Continental Drift, chap. 5. New York: Academic Press.

WILSON, J. T. (1963). Continental drift. Sci. Am., 208, no. 4.

WILSON, R. L. (1962). The palaeomagnetism of baked contact rocks and reversal of the earth's magnetic field. Geophys. J., 7: 194–201.

PRINCIPLES OF TIME-STRATIGRAPHIC CLASSIFICATION IN THE PRECAMBRIAN

C. H. Stockwell, F.R.S.C.

ABSTRACT

Time-stratigraphic units should be based on recognizable geological events and, for convenience of description and discussion, the units should be named. The upper and lower limits of each unit should be defined in terms of rock in a type area or region which serves as a standard for the unit. Orogenies and major unconformities serve as the basis for division of Precambrian time into major units, of eon and era rank, and those orogenies that are geographically the most extensive serve best. Isotopic ages are subject to various interpretations and should not be used for definition of time units but are extremely useful in long-range correlation. These and other principles are applied to the Canadian Shield as an example.

IN THE PRECAMBRIAN, time relationships are determined mainly by superposition of strata, and intrusive and metamorphic relationships combined with isotopic dates. Before the advent of isotopic dating techniques, correlations were based on such criteria as lithological similarity, comparison of sequences, and relation to adjacent strata, to unconformities, and to intrusions. Such features are still useful, especially for short-range correlations, but it is now possible to make interregional and world-wide correlations with a considerable degree of confidence. Isotopic dating techniques, however, have brought with their great advantages certain problems both in the interpretation of the dates themselves and also in how best they may be used in building up a time classification of Precambrian rocks. Solutions to these problems are proposed, mainly on the basis of experience in the Canadian Shield.

Geology is not an exact science and a calculated isotopic age is subject to interpretation in terms of geological factors which may have brought about changes in the normal ratio of decay products. For example, if the daughter element is lost subsequent to the time of crystallization of a mineral that is being studied, the calculated age gives the time at which the loss occurred and is younger than the mineral. Should the mineral become enriched in the daughter element by contamination, either at the time of crystallization or by later introduction, the measured age is too old. It is only under conditions not affecting the normal ratio that the true age is obtained.

Examples in which anomalous dates have been found are well known. One is the case where a dyke or other intrusive body gives an age that is significantly older than that obtained on the rock which it cuts. Another is the case where two coeval mineral species give widely different ages by the same isotopic method, or where the same mineral sample gives different dates by different methods. It is also common to find that rocks known to differ widely in age give practically the same calculated ages. All these and other examples require explanation in terms of the geological conditions that brought about the discrepancies. Moreover, it is apparent that any number of checks by the same method on the same mineral from different parts of a body being studied give no assurance that the age obtained is reliable, for all could be affected by some constant factor of error.

The most convincing isotopic evidence for the reliability of an age is found by obtaining cross-checks by different methods on the same material or by the same method on different mineral species that are coeval. Agreement indicates that the isotopic ratios are normal, for if they had been affected by extraneous geological factors it would be extremely unlikely that entirely different isotopes would behave in exactly the same way. By making tests of this nature it has also been found that some methods and some minerals give more reliable results than others.

Another troublesome factor is the analytical error, which may be more than ± 100 m.y. in Archaean rocks. The margin of error may be reduced by averaging a number of determinations or, in some instances, by considering field relations.

Experience in the Canadian Shield has shown that ages that are clearly discrepant, well beyond the analytical error, are found mostly in unusual geological settings such as near boundaries between orogens of widely different ages. Elsewhere, over huge areas, reasonably good checks have been obtained on different minerals and by different methods although much work still remains to be done. Agreement with the known succession of geological events is found most commonly, and it is chiefly for this reason that the great majority of the dates are considered to be reasonably reliable. However, there is always some degree of uncertainty and all require interpretation to a greater or less degree.

For the purpose of building up a time-stratigraphic classification of Precambrian rocks the uncertainties in interpretation and in their degree of accuracy must be taken into account and any classification based solely on isotope dates would be unsound. It is recommended instead that the fundamental principles proposed by the American Commission on Stratigraphic Nomenclature (1961) be followed, namely that time units be based on actual rock rather than abstract time and that a type locality be used for definition of the unit. Isotopic dates then become useful, as are fossils, for the purpose of correlation. Even if the true age of the rock at the type locality is not precisely known, the geologist may, in his best judgment, correlate other rocks with it. The type locality serves as a standard for reference and any mistakes made in correlation do not invalidate the unit. Likewise any improvements that may be made in methods of dating or any changes made in interpretation do not affect its validity. In the Precambrian, the rocks forming the boundaries of the unit should represent some recognizable and significant geological event and should be suitable for dating by isotopic methods. For convenience of description and discussion the formally defined units should be given names.

As stated by the American Commission, the principal purposes of time-stratigraphic classification are for correlation and for placing rocks in a systematic chronological sequence. Accordingly, when these purposes are not served it is pointless to use time terms such as system or series; in such instances rock-stratigraphic terminology is preferable. In any case, in the Precambrian, it is generally impracticable to date sedimentary systems and series isotopically if these terms are used in the sense in which they are used in the Phanerozoic, where they represent the interval of time that extended from the beginning to the end of deposition.

Time-stratigraphic classification is concerned mainly with sedimentary and extrusive rocks, but is applied also to intrusive igneous rocks. Igneous rocks in which the normal ratio of decay products has not been altered may be dated directly by isotopic means. They may for all, practical purposes follow a single time horizon or, if related intrusions are grouped together, the time unit may be defined between upper and lower time limits. Sedimentary rocks, in the Precambrian, cannot ordinarily be dated directly and difficulties are consequently encountered in their correlation. This contrasts with the situation in the Phanerozoic, where sedimentary rocks are correlated by means of the fossils contained in them. However, some sedimentary assemblages in the Precambrian may be dated approximately by using suitable interlayered volcanic material. They may also be bracketed between maximum and minimum age limits by their relation to other rocks which are suitable for dating. The maximum is given by the age of unconformably underlying igneous and metamorphic rocks or by detrital minerals within the sediments. The minimum is given by overriding metamorphism and by intrusive rocks and other transecting features. Orogenies serve very well for defining the maximum and minimum limits of large time-stratigraphic units in situations where a sedimentary assemblage unconformably overlies an older orogen and is itself involved in a younger orogeny. By virtue of the plutonic and metamorphic rock contained in them both orogenies may be well suited for isotopic dating, but the dating of the older one becomes more difficult when it has everywhere been overridden by the younger. Large time-units between maximum and minimum limits may contain geographically separated assemblages of sedimentary and volcanic rocks that are not necessarily of the same age. They also contain hiatuses.

Those boundaries of a time-stratigraphic unit that can be recognized in the field over the greatest geographical extent serve best for definition of the unit. Beyond such recognizable limits the time boundaries may pass through a conformable succession and be unrecognizable by field methods.

In some such cases, however, they may be located approximately by the isotopic dating of flows, dykes, sills, or other features not precisely equivalent in age to that of the time boundary sought. This difficulty is unavoidable, but the method of defining the units by significant observable features at type localities is still preferable to using isotopic methods alone because the boundaries would then nowhere coincide with objective geological features.

First-order time-stratigraphic subdivisions of the Precambrian should be few in number, preferably only two. Such a numerically simple subdivision has certain advantages. One is that large units remain available for comparing or contrasting gross features of geological history. For example, it may be desired to compare or contrast certain features of the Precambrian with those of the Phanerozoic, or of the Archaean with those of the Proterozoic. Another advantage is that large units become useful where age data are incomplete. For example, many rock units can be classified as Archaean by means of only the minimum age whereas, if this large first-order unit were not available but were replaced by two first-order units, the rock units could not be classed under either of them. Similarly, second-order subdivisions of each of the major units should also be few in number.

Units of equal rank need not necessarily span, even approximately, the same length of time but should be selected to set apart those natural units that are the most serviceable over the largest area.

It is most practical to define the first-order units first and to make subdivisions later as more detailed information becomes available. This contrasts with the procedure followed in the Phanerozoic, where systems are the fundamental unit and the next larger unit is formed by a combination of systems.

Orogenies and tectonic cycles are not time-stratigraphic units. An orogeny is a period of mountain-building commonly accompanied by folding, metamorphism, and the emplacement of plutonic rocks. A tectonic cycle has larger scope. It begins with the geosynclinal deposition of sedimentary and volcanic materials, culminates in an orogeny, and ends with the deposition of sediments derived from the mountains. Such cycles are recognizable in the Precambrian but, on the other hand, some orogenies have overridden an older orogen without an intermediate geosynclinal phase of deposition. For the purpose of subdivision of Precambrian time, tectonic cycles are therefore of more limited value than are time-stratigraphic units. It is also difficult to date the time of beginning and end of deposition of the sedimentary rocks which define the limits of a cycle.

The proposed principles of time-classification in the Precambrian may be illustrated by their application in the Canadian Shield. Some 450 potassium-argon determinations in the Shield have now been made by the Geological Survey of Canada (Lowdon 1960, 1961, 1963, and Leech *et al.* 1963). In addition, many dates have been reported by other workers, among whom may be mentioned Beall (1960, 1961), Burwash and Baadsgaard (1962*a, b*), Fairbairn *et al.* (1960), and Goldich *et al.* (1961). For details on the classification of rocks of the Shield the reader is referred to previous reports (Stockwell 1961, 1962, 1963a, 1963b).

The Shield is divided into seven main structural provinces and several subprovinces (Fig. 1). These serve as objective divisions for reference; they have geographical boundaries and are not defined by age but contain rocks of various ages. The isotopic ages determined in each province reflect mainly the last important orogeny which produced the structural features which characterize the province, although some of the isotopic dates reflect both older and younger events that took place within each province. Orogenies, in conjunction with major unconformities and isotopic age determinations, serve to define the maximum and minimum limits of major time-stratigraphic and corresponding geologic time units.

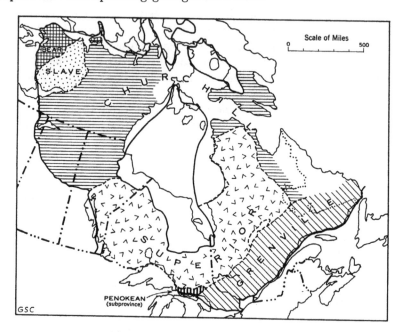

FIGURE 1. Main structural provinces of the Canadian Shield.

Three main orogenies have been distinguished, each having affected very large areas. These are named, respectively, the Kenoran, the Hudsonian, and the Grenville orogenies. The type region for the Kenoran is the Superior Province and it is defined as the last major period of folding, metamorphism, and plutonic intrusions that affected that province. Its potassium–argon age, which is not part of the definition, has a peak at 2500 m.y. and, taking the analytical error into account, has a probable duration of 100 m.y. or thereabouts. On the basis of similarities in age and geological relationships the last orogeny in the southern part of the Slave Province is correlated with the Kenoran. The type region for the Hudsonian orogeny is the Churchill Province, and the potassium–argon determinations have a peak at 1700 m.y. with a spread similar to that of the Kenoran. The Penokean Subprovince and the Bear Province were involved in an orogeny correlated with the Hudsonian. Likewise, the type region for the Grenville orogeny is the Grenville Province. The potassium–argon ages on biotite average 910 m.y., those on muscovite average 950 m.y., and a few dates on uraninite and thorianite average 980 m.y. An orogeny of similar age in the Appalachian Province is correlated with the Grenville. The Duluth gabbro of approximately equivalent age serves to extend the time horizon beyond the limits of the Grenville orogen.

The Kenoran in its type region forms the basis for a division of all Precambrian time into two eons called the Archaean, with a K–Ar age of 2450 m.y. and older, and the unconformably overlying Proterozoic spanning the range from 2450 m.y. to the base of the Cambrian at about 600 m.y. Important stratigraphic and plutonic subdivisions of the Archean are recognized locally by field relations, but these should not be given formal time-stratigraphic names until such time as they will serve the purpose of interregional correlation.

The Hudsonian and Grenville orogenies form the basis for a subdivision of the Proterozoic into three eras called the Lower (Early) Proterozoic, 2450 to 1650 m.y., the Middle Proterozoic, 1650 to 900 m.y., and the Upper (Late) Proterozoic, 900 to about 600 m.y. The sedimentary and volcanic rocks of each of the three eras are separated from underlying orogens by a major unconformity. The Upper Proterozoic rocks of the Shield have not been involved in an orogeny, but the Lower and Middle Proterozoic each contain orogenic plutonic rocks in addition to the sedimentary and volcanic materials. Thus, two subdivisions of these two eras are indicated, the plutonic rocks mainly forming a late stage and the sedimentary and volcanic rocks mainly an early stage, but as the time line between the two is indefinite they

TABLE I

TABLE OF FORMATIONS OF THE CANADIAN SHIELD

Eon	Era	Rock unit
Proterozoic	Upper (Late) Proterozoic	Upper Keweenawan, Double Mer (may be younger)
	Middle Proterozoic	Duluth Gabbro, Muskox intrusion, Logan intrusions, Killarney Granite; Lower and Middle Keweenawan, Hornby Bay, Athabasca, Sims
	Lower (Early) Proterozoic	Sudbury irruptive, Nipissing Diabase, Metachewan Diabase, St. Cloud and other granites; Huronian, Animikie, Kaniapiskau, Snare, Great Slave, Nonacho
Archaean		Prosperous, Preissac-Lacorne, Giants Range, Vermilion, Saganaga, and other granites; Keewatin, Abitibi, Coutchiching, Pontiac, Timiskaming, Seine, Knife Lake, Ridout, Windigocan, Sickle, Rice Lake, Yellowknife

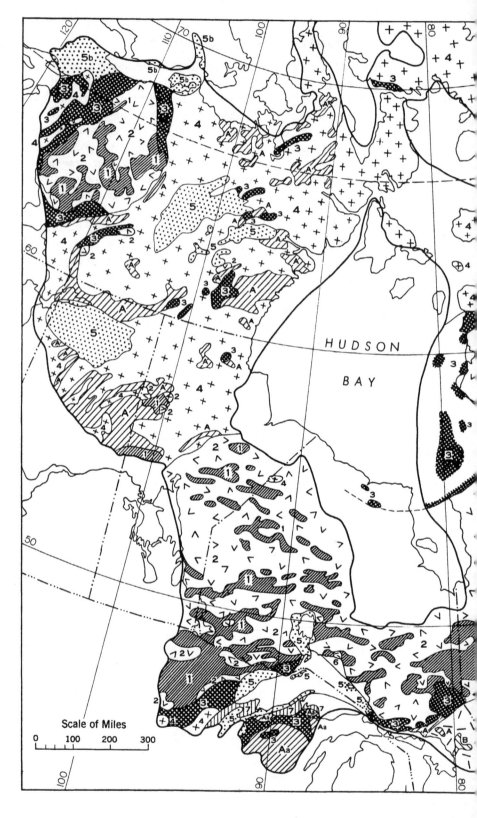

FIGURE 2. Geological ske

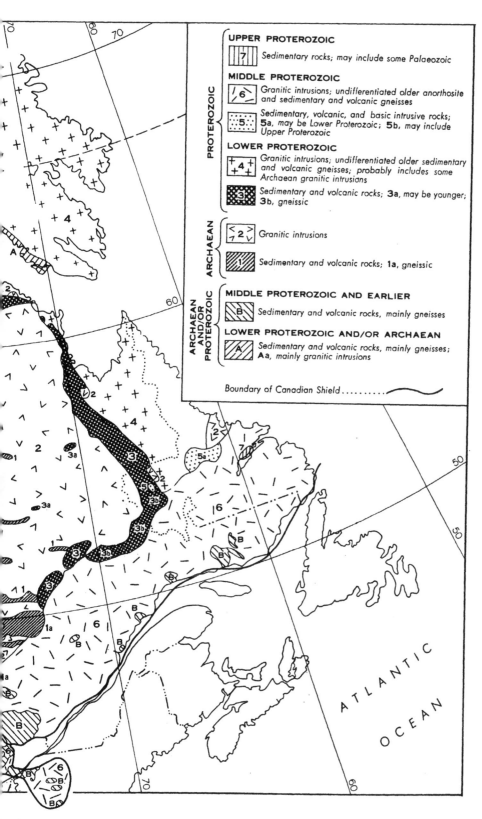

UPPER PROTEROZOIC

| 7 | Sedimentary rocks; may include some Palaeozoic

MIDDLE PROTEROZOIC

| 6 | Granitic intrusions; undifferentiated older anorthosite and sedimentary and volcanic gneisses

| 5 | Sedimentary, volcanic, and basic intrusive rocks; 5a, may be Lower Proterozoic; 5b, may include Upper Proterozoic

LOWER PROTEROZOIC

| 4 | Granitic intrusions; undifferentiated older sedimentary and volcanic gneisses; probably includes some Archaean granitic intrusions

| 3 | Sedimentary and volcanic rocks; 3a, may be younger; 3b, gneissic

| 2 | Granitic intrusions

| 1 | Sedimentary and volcanic rocks; 1a, gneissic

MIDDLE PROTEROZOIC AND EARLIER

| B | Sedimentary and volcanic rocks, mainly gneisses

LOWER PROTEROZOIC AND/OR ARCHAEAN

| A | Sedimentary and volcanic rocks, mainly gneisses; Aa, mainly granitic intrusions

Boundary of Canadian Shield.........

ᵖ of the Canadian Shield.

have not been named. Named rock units that have so far been placed within the units of eon and era rank are shown in Table I.

A geological sketch map of the Canadian Shield is given in Figure 2. On this map most of the rock units have been assigned to age categories on the basis of field relations in combination with isotopic dates. Some have not yet been well dated and are assigned to age categories mainly on the basis of geological considerations. The reader may compare this geological map with the tectonic map of the Shield published previously (Stockwell 1962).

REFERENCES

AMERICAN COMMISSION ON STRATIGRAPHIC NOMENCLATURE (1961). Code of Strati-graphic Nomenclature. Bull. Am. Assoc. Petrol. Geol., *45*: 645–65.

BEALL, G. H., SAUVE P., and staff (1960). Age investigations in New Quebec and Labrador. N.Y.O. 3941, Eighth Ann. Progr. Reprt. for 1960, Dept. Geol. and Geophys., Mass. Inst. Technol., pp. 245–53.

—————— (1961). Age investigations in Ungava, New Quebec. N.Y.O. 3942, Ninth Ann. Progr. Reprt. for 1961, Dept. Geol. and Geophys., Mass. Inst. Technol., pp. 193–9.

BURWASH, R. A. and BAADSGAARD, H. (1962a). Precambrian K–Ar dates from the Western Canada Sedimentary Basin. J. Geophys. Res. 67: 1617–25.

—————— (1962b). Yellowknife–Nonacho age and structural relations. *In* The Tectonics of the Canadian Shield, *edited by* J. S. Stevenson (Roy. Soc. Can., Spec. Publ. no. 4), pp. 22–29. Toronto: University of Toronto Press.

FAIRBAIRN, H. W., HURLEY, P. M., and PINSON, W. H. (1960). Mineral and rock ages at Sudbury–Blind River, Ontario. N.Y.O. 3941, Eighth Ann. Progr. Rept. for 1960, Dept. Geol. and Geophys., Mass. Inst. Technol., pp. 7–42.

GOLDICH, S. S., NIER, A. O., BAADSGAARD, H., HOFFMAN, J. H., and KRUEGER, H. W., (1961). The Precambrian geology and geochronology of Minnesota. Univ. Minne-sota, Minnesota Geol. Surv., Bull. 41.

LEECH, G. B., LOWDON, J. A., STOCKWELL, C. H., and WANLESS, R. K. (1963). Age determinations and geological studies. Geol. Surv. Can., Paper 63–17, pp. 5–121.

LOWDON, J. A. (Compiler) (1960). Age determinations by the Geological Survey of Canada. Geol. Surv. Can., Paper 60–17, pp. 5–40.

—————— (1961. Age determinations by the Geological Survey of Canada. Geol. Surv. Can., Paper 61–17, pp. 5–86.

—————— (1963). Age determinations by the Geological Survey of Canada. Geol. Surv. Can., Paper 62–17, pp. 5–120.

STOCKWELL, C. H. (1961). Structural provinces, orogenies, and time classification of rocks of the Canadian Precambrian Shield. *In* Geol. Surv. Can., Paper 61–17, pp. 108–18.

—————— (1962). A tectonic map of the Canadian Shield. *In* The Tectonics of the Cana-dian Shield, *edited by* J. S. Stevenson (Roy. Soc. Can., Spec. Publ. no. 4), pp. 6–15. Toronto: University of Toronto Press.

—————— (1963a). Second report on structural provinces, orogenies, and time classification of rocks of the Canadian Precambrian Shield. *In* Geol. Surv. Can., Paper 62–17, pp. 123–33.

—————— (1963b). Third report on structural provinces, orogenies, and time classification of rocks of the Canadian Precambrian Shield. *In* Geol. Surv. Can., Paper 63–17, pp. 125–131.

AGE AND CORRELATION PROBLEMS IN THE APPALACHIAN REGION OF CANADA

W. H. Poole, D. G. Kelley, and E. R. W. Neale

ABSTRACT

The problem of establishing the age of strata and their correlations is reviewed. The region is the most intensely studied tectonic province in Canada.

Classical methods of palaeontological dating and of extrapolation and correlation by like lithology, similar sequences, unconformities, and alleged synchroneity of intrusions have been more successful in some places than in others. The abundance of fossils varies greatly from one system to another and within individual systems. Thick Ordovician and Silurian formations with few fossil localities in much of central New Brunswick, southern Nova Scotia, and central Newfoundland are correlated by the other classical methods less dependent on palaeontology. These methods are an undesirable means of placing time-surfaces and require constant review as new data become available.

Unconformities have served as time-surfaces. Unfortunately, most are not precisely dated and are not equally well developed throughout the region. Similarly, ultramafic and granitic bodies are not precisely dated stratigraphically but are known to have been intruded at different times.

In Carboniferous stratigraphy, present rock "groups" are actually time-stratigraphic units. The definition of true rock-stratigraphic units is required to preclude the need for formational boundaries based solely on fossils where no lithological change is apparent, and to emphasize the significance of the transgressive nature of some formations.

Absolute age determination studies by K–Ar and Rb–Sr methods have indicated correlations of granitic intrusions and metamorphism throughout the region. Initial studies by the Massachusetts Institute of Technology yielded a minimum age of Lower Devonian strata, a contribution to calibrating the time-scale between relative and isotopic ages.

K–Ar age studies by the Geological Survey of Canada supplemented the initial M.I.T. work and broadened the coverage. Bar diagrams of nearly all determinations show distinct concentrations of dates of Precambrian (Grenville, 875–975 m.y.), Ordovician (Taconic, 475–500 m.y.), Devonian (Acadian, 350–400 m.y.), and Cretaceous (Monteregian, 100–125 m.y.). Ultramafic bodies were first emplaced in the Ordovician. Most Palaeozoic granitic bodies are mid-Devonian, a few are Ordovician, and two are Carboniferous.

Approaches to interpreting anomalously low and high dates are reviewed. No K–Ar dates have been obtained that are older than those indicated by the geological relations within the limits of analytical error for a single determination. Most cases of younger ages are explained by loss of argon through superimposed younger thermal events.

The Rb–Sr whole-rock isochron method appears likely to yield more useful results because the isotopic ratios apparently remain substantially unchanged in the sample through weak metamorphic events, events strong enough to change K–Ar ratios.

Brief palaeomagnetic studies in the region are limited to Silurian strata in northern Newfoundland, to Carboniferous strata in most provinces, and to Monteregian plugs in southern Quebec.

No intractible conflicts exist between ages provided by palaeontological methods and the newer methods, isotopic and palaeomagnetic. All methods augment and supplement each other to the continuing improvement of the synthesis of the tectonic history of the region.

IN THE Appalachian region of Canada, as everywhere else, the fundamental objective of geological study is the qualitative and quantitative understanding of the region in terms of time, biological processes, lithology, and geometry. Time is all-important, and surfaces of equi-time are an essential part of any model that purports to illustrate the history of a region. The passage of geological time is recorded in the nature and thickness of sedimentary rocks, the superposition of bed upon bed, the nature of the breaks between beds, and the intrusive relationship between sedimentary and igneous rocks. Geological time is measured as fixed points in the evolution of former plants and animals, by the increasing concentration of daughter elements formed during the radioactive decay of certain unstable isotopes, and by the apparent wandering of the earth's magnetic poles.

Much material basic to these concepts was derived from early studies of the Appalachian mountain system in the United States. The early work of Logan, Dawson, and others in the Appalachian region of Canada kept pace with that in other parts of the mountain system as new facts were noted and new hypotheses were tested. This work was continued by their successors, and the region is now the most thoroughly mapped and studied geological subdivision of Canada. Whether or not its history is best understood is a moot point. Nevertheless, this is an appropriate opportunity to appraise the attempts to establish time-planes within the rocks themselves, as new models and interpretations of the geology of the Appalachian region of Canada are constructed.

Many stratigraphic sections have been measured and described in detail, and their relative ages have been determined both by superposition and by the fauna and flora contained in them. Recently, the examination of microfauna and microflora has aided the correlation of some strata that lack definitive megafossils and has increased the precision of dating by megafossils of other strata. The *bête noire* of stratigraphy, the distinction between time-stratigraphic, biostratigraphic, and rock-stratigraphic units, has insinuated itself into the Canadian Appalachian scene, as elsewhere, and revisions and redefinitions of established stratigraphic units are commencing.

Isotopic dating of the inorganic parts of rocks has been a great aid in understanding the metamorphic and intrusive history of the region and in the assigning of numerical limits to the relative ages determined by fossils. During the early stages of study, dates obtained from some granitic rocks by the K–Ar and Rb–Sr methods substantiated the ages that had been deduced by their relationships to rocks dated by standard palaeontological and stratigraphic methods. However, some of the dates obtained in the past few years do not fit so neatly into this historical framework. Their significance will be examined in this report.

Palaeomagnetic studies are still in the experimental stage in the Appalachian region of Canada. Only a few attempts have been made to determine absolute ages on the basis of palaeo-pole positions. These studies have provided a useful tool in the correlation of certain types of rock in a few localities.

This brief examination of classical, isotopic, and palaeomagnetic methods of correlation and age determination in the region does not pretend to be comprehensive. The authors have drawn chiefly on examples from areas with which they are familiar.

CLASSICAL METHODS OF CORRELATION AND THEIR PROBLEMS

Early workers in the Appalachian region of Canada soon identified four first-order stratigraphic and tectonic divisions: Precambrian, Early and Middle Palaeozoic, Permo-Carboniferous, and Triassic (Figs. 1 and 4). Lithology, stratigraphic succession, structure, and fossils have been used to recognize and define divisions and then to extrapolate them from the type localities to lesser-known areas.

Density of Fossil Localities

Fossils, the key to relative age, vary greatly in abundance not only from one primary division to another but also within individual divisions (Fig. 2). Thus fossil localities are abundant in most Silurian and Devonian strata of Gaspé and adjacent New Brunswick and in almost all Carboniferous strata, i.e., sufficiently abundant to identify successive epochs of geological time. Fossil localities are less abundant in the Silurian and Devonian strata of western New Brunswick, adjacent Quebec, and southern New Brunswick; in the Cambrian and Ordovician strata of western and northwestern Newfoundland; and in the Ordovician and Silurian strata along the coast of Notre Dame Bay, northern Newfoundland. In these regions the combination of fossils and lithology permits fairly accurate delineation of the rocks that were deposited during any one period of geological time, but finer subdivisions are possible only in local areas. Fossil localities are rare or lacking in the Ordovician to Lower Devonian rocks of central New Brunswick, southern Nova Scotia, and central Newfoundland. These areas contain thick geosynclinal assemblages mainly of slate, greywacke, chert, and volcanic rocks, which are generally more highly deformed, metamorphosed, and intruded than correlative strata elsewhere in the region. Some fossils have been obliterated by these processes, but most of these rocks never did contain abundant organic remains. Such rocks will long continue to present problems of correlation.

Success in correlating time-stratigraphic units depends, of course, on the range of the fossils present as well as their abundance. Faunal zonation has been established for several detailed sections of Lower Palaeozoic rocks

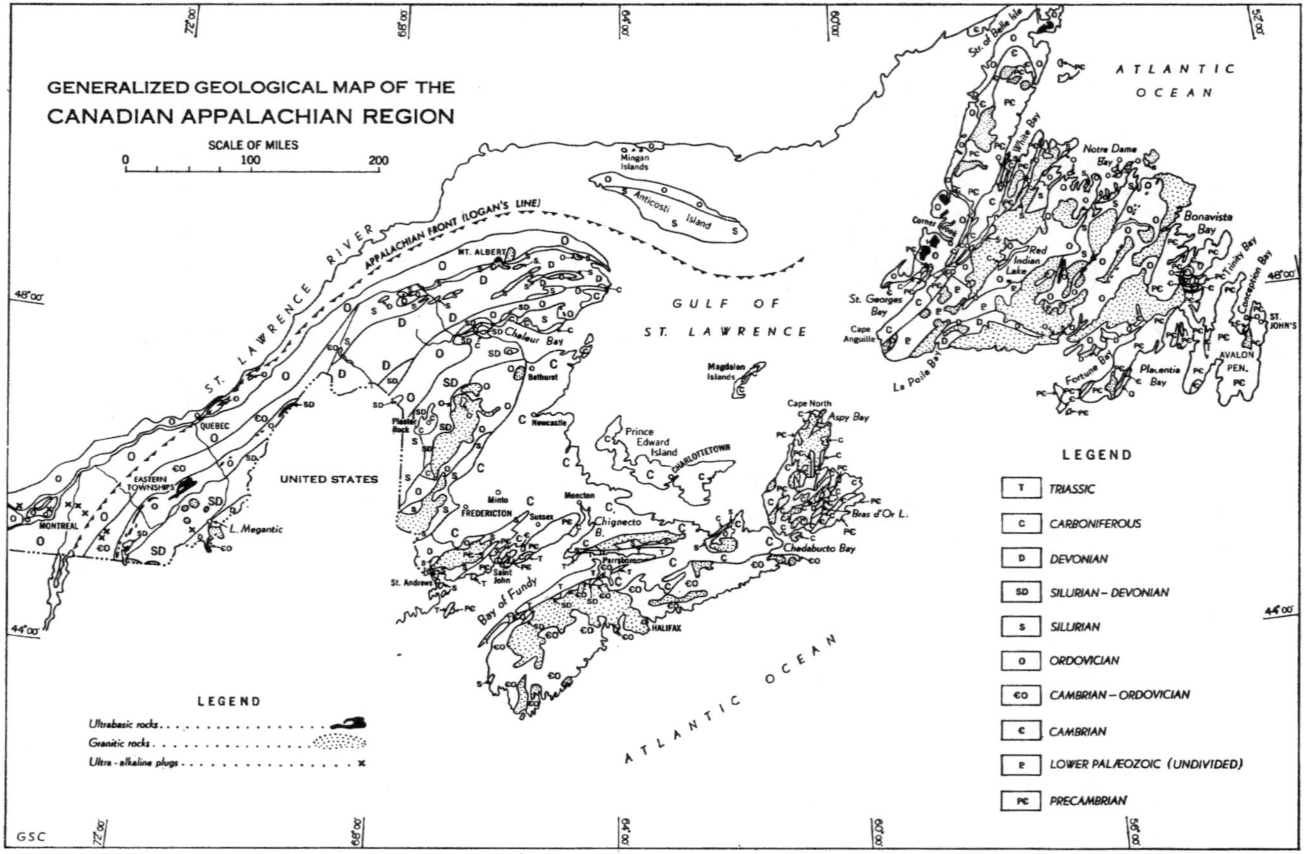
GENERALIZED GEOLOGICAL MAP OF THE
CANADIAN APPALACHIAN REGION
SCALE OF MILES
0
100
200
ST. LAWRENCE RIVER
APPALACHIAN FRONT (LOGAN'S LINE)
GULF OF ST. LAWRENCE
ATLANTIC OCEAN
UNITED STATES
MONTREAL
QUEBEC
EASTERN TOWNSHIPS
L. Megantic
MT. ALBERT
Chaleur Bay
Bathurst
Newcastle
Plaster Rock
FREDERICTON
Minto
Moncton
Sussex
Saint John
St. Andrews
Bay of Fundy
Chignecto B.
Parrsboro
HALIFAX
Chedabucto Bay
Prince Edward Island
CHARLOTTETOWN
Magdalen Islands
Mingan Islands
Anticosti Island
Cape North
Aspy Bay
Bras d'Or L.
St. Georges Bay
Cape Anguille
La Poile Bay
Corner Brook
Red Indian Lake
White Bay
Str. of Belle Isle
Notre Dame Bay
Bonavista Bay
Trinity Bay
Conception Bay
ST. JOHN'S
AVALON PEN.
Placentia Bay
Fortune Bay
72°00'
68°00'
64°00'
60°00'
56°00'
52°00'
48°00'
44°00'
LEGEND
Ultrabasic rocks
Granitic rocks
Ultra-alkaline plugs
LEGEND
T TRIASSIC
C CARBONIFEROUS
D DEVONIAN
SD SILURIAN-DEVONIAN
S SILURIAN
O ORDOVICIAN
€O CAMBRIAN-ORDOVICIAN
€ CAMBRIAN
Ƥ LOWER PALÆOZOIC (UNDIVIDED)
P€ PRECAMBRIAN
GSC

FIGURE 1

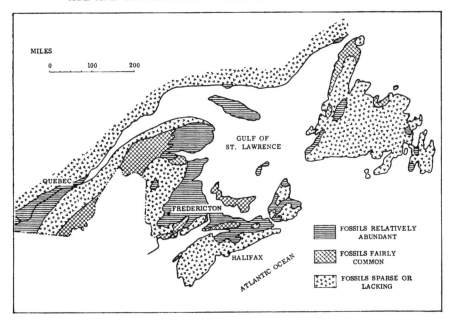

FIGURE 2.   Relative densities of recorded fossil localities in Palaeozoic rocks.

and for many sections of Carboniferous rocks. Where fossil localities are few, it matters little how carefully the position of particular fossils has been located in the geological column—great gaps in the historical record still exist, and attempts to fill them must rely on large extrapolations from the few well-known areas. These extrapolations are based on lithology, structural breaks, intrusive history, and, to a lesser extent, isotopic age determinations.

Problems of Lithological Correlations

Correlation, without fossils, is necessarily by similarities in lithology and lithological sequences and is usually less than satisfactory. Nevertheless, large terranes throughout the Canadian Appalachian region are presently dated by extrapolations from only one or two fossil localities.

A good example is the Meguma Group of Nova Scotia. For many years, the Meguma was interpreted as Precambrian because of its presumed lithological similarity to known Precambrian rocks of southeastern Newfoundland (Woodman 1908; Alcock 1947). Graptolites, found in slates of this group about 10 years ago by Crosby (1952, 1963) and subsequently by Smitheringale (1960), had actually first been reported by Ami (1903). They have been identified as Early Ordovician forms by O. M. B. Bulman (Crosby 1963), and it is now the fashion to refer to the entire Meguma Group as "Lower Ordovician," "Cambro-Ordovician," or "Lower Ordovician and Older." However, the few fossil localities are restricted to a small area in the northwestern boundary zone of the group. Furthermore, these graptolitic slates represent the uppermost part of a great thickness (about 30,000 feet) of slates and quartzites (F. C. Taylor, pers. comm. 1963). It is, therefore, still conceivable that the lower part of the group is Eo-Cambrian and, until the chance discovery of more fossils or the development of new isotopic or other methods capable of determining the age of sedimentation, the age of the lower part of the Meguma will remain unknown.

In some cases, long correlations from a single fossil locality are borne out by subsequent work. Thus, the Tetagouche Group, a thick sedimentary-volcanic complex in northern New Brunswick, was dated as Middle Ordovician on the basis of a single graptolite locality (Smith 1957). An additional Ordovician shelly fossil locality has now been discovered in the region (L. M. Cumming, pers. comm. 1962), and several localities of definite and probable Middle Ordovician graptolites and shells have been found recently in the southwestward extension of the group (Poole 1960, 1963; Anderson and Poole 1959).

More commonly, however, new fossil finds have emphasized the hazards of lithological correlations. The Middle Silurian greywacke and sparsely graptolitic slates of central New Brunswick (Anderson and Poole 1959) can be traced with only minor variations into the Charlotte Group of southernmost New Brunswick. Until 1962, no fossil localities were known in this group but, as it was overlain unconformably by Middle Silurian strata, it was assumed to be Ordovician (Alcock 1947). At least part of the group is now known to contain Early Ordovician graptolites (L. M. Cumming, pers. comm. 1962). Were it not for the clearly unconformable relations with the Silurian, the Charlotte Group may well have been considered part of the Middle Silurian terrane.

Major revisions in the synthesis of the geological history of the region must be expected where correlations of thick assemblages are based on lithology. Such correlations presently play a large part in interpretations of the geology of central Newfoundland.

Problems of Correlating by Unconformities and Intrusions

Consciously or unconsciously, geologists have used unconformities as time-surfaces in gross correlations throughout the relatively unfossiliferous terranes of the Appalachian region of Canada. Early workers noted angular unconformities between Precambrian basement rocks and overlying, less-deformed, Cambrian (and Eo-Cambrian) strata. Unconformities between Ordovician and Silurian strata were also recognized in several places, particularly in southern Quebec, and were correlated with the Taconic orogeny of Middle or Late Ordovician age for which evidence had first been recorded in northern New York and Vermont. Carboniferous rocks, and in one locality Upper Devonian rocks, lie with marked angular unconformity on all older rocks, and commonly contain pebbles of the granites which intrude these older rocks, thus providing the main evidence of the Acadian orogeny in mid-Devonian time.

Ultramafic rocks in the Canadian Appalachian region are now generally considered to have been first emplaced during the Ordovician (Hess 1939) Taconic orogeny. Some workers have considered them as probably Devonian (e. g. McGerrigle 1954; Marleau 1959), but Neale (1957) showed in northern Newfoundland that serpentinite bodies which cut post-Ordovician rocks had been remobilized and reintruded.

Granite bodies cut Lower Devonian rocks in northern New Brunswick (Alcock 1941), Nova Scotia (Crosby 1963), and Newfoundland (Cooper 1954). Hence, all granitic rocks of the region, except known Precambrian granites of the basement complex, were assumed to be mid-Devonian and related to the Acadian orogeny (Weeks 1957).

It has been natural, then, to use unconformities or supposed unconformities and the intrusive rocks as time-markers in interpreting the history of unfossiliferous areas—particularly those which are known only through reconnaissance studies. Thus, in central Newfoundland, the pillow lava–slate–greywacke assemblages which contain ultramafic intrusions are assumed to be Ordovician and to have been deformed during the Taconic orogeny. Less metamorphosed, volcanic-sedimentary assemblages (which resemble fossiliferous Silurian assemblages of the coastal areas) are presumed to be post-Taconic and separated from the lava–slate–greywacke sequence by an unconformity. Granitic rocks which intrude both sequences are interpreted as Devonian and related to the climactic Acadian orogeny. This is an extreme example which, supplemented by a few isotopic age determinations, was used by two of the present authors in construction of a tectonic map of the Canadian Appalachian region (Neale, Béland, Potter, and Poole 1961). This reasoning is common to practically all interpretations of the Lower and Middle Palaeozoic mobile belt and must be used in the absence of fossils. Revisions are made continually as newly discovered data are applied. However, the broad picture drawn in Newfoundland by these methods has accommodated the additional data gathered over the past few years.

The pitfalls, however, are many and some of them are just in the process of discovery. The angular unconformities within the Lower Palaeozoic rocks are nowhere precisely dated. Thus, the Taconic unconformity in southern Quebec consists of definite and probable Middle Ordovician strata overlain by Silurian strata; and in southern New Brunswick it consists of Lower Ordovician strata overlain by Middle Silurian strata. More important, growing evidence suggests that the deformational effects of Taconic orogeny varied from place to place across the region, probably from one northeasterly striking belt to the next. In Gaspé, Ordovician and Silurian strata lie in contact along an angular unconformity (Béland 1960); in northern New Brunswick, the contact is in some places faulted and in others unconformable (Smith 1957; Greiner 1960); in central New Brunswick, it is either faulted and/or structurally conformable (Poole 1960, 1963); in southern New Brunswick, it is an angular unconformity (Alcock 1960); and, finally, in northwestern Nova Scotia, strata of three systems—Ordovician, Silurian, and Devonian—are mainly structurally conformable (Smitheringale 1960; Taylor 1962). In Newfoundland, along the western part of Notre Dame Bay, probable Silurian strata rest unconformably on Ordovician strata, which are cut by granitic rocks (Neale and Nash 1963), whereas in the eastern part of the bay, about 50 miles away, the Ordovician and Silurian strata are conformable, a relation originally described by Twenhofel and Shrock (1937) and recently confirmed by Williams (1962, 1963). Unconformities are not everywhere equally well developed or easily recognized, and caution must accompany their use as approximate equi-time surfaces.

Some Palaeozoic granitic rocks are now known to be older and some younger than the mid-Devonian granites which form the bulk of the Palaeozoic granites in the Appalachian region of Canada. Two years ago (Neale, Béland, Potter, and Poole 1961), no Ordovician granites had been identified positively, although their presence was suspected because Ordovician granites had been mapped in New Hampshire (Billings 1956), and two anomalously old isotopic ages had been obtained from Cape Breton Island (Fairbairn, Hurley, Pinson, and Cormier 1960). Ordovician granitic bodies, large and small, have now been identified by stratigraphic evidence in the Eastern Townships (Duquette 1960), and by stratigraphic evidence supplemented by isotopic age determinations within the ultramafic bodies of southern Quebec (Poole, Béland, and Wanless 1963), within gneiss in central New Brunswick (Poole 1963), and in the western part of Notre Dame Bay, Newfoundland (Neale and Nash 1963).

Only two granitic bodies of the Appalachian region of Canada are known by stratigraphic evidence to be younger than the mid-Devonian granites although such young granites are abundant in Rhode Island, U.S.A. (Hurley, Fairbairn, Pinson, and Faure 1959). Sills and plugs cut the Upper Mississippian Mispek Group near Saint John, southern New Brunswick (Alcock 1938); and the Belleoram granite pluton cuts fossiliferous Upper Devonian conglomerates in Fortune Bay, southeastern Newfoundland (Weeks 1957). Hence, caution must be exercised in any given area when assuming that all the granites are mid-Devonian.

Problems of Rock and Time-Rock Units

Detailed faunal studies have been carried out and are still progressing on well-exposed sections of all the Lower and Middle Palaeozoic systems of the region. Biozonal divisions have been established in parts of the Cambrian (e.g. Hutchinson 1952, 1963), Ordovician (e.g. Berry 1962), Silurian (e.g. McLearn 1924; Northrop 1939), and Devonian (Boucot and Cumming 1963). Apart from detailed studies of this nature, however, mapping of pre-Carboniferous strata has resulted in rock-stratigraphic units. Fossils have been used only to determine the age of the unit, and to place it in the stratigraphic column. Some workers have attempted to define formations on the basis of faunal zones, in contravention of the Code of Stratigraphic Nomenclature (Amer. Comm. on Strat. Nomen. 1961), but these attempts are few, possibly because fossils are generally scarce. However, fossils have been used commonly to identify formations in Carboniferous strata. Fossil localities are relatively abundant, and detailed studies of flora and fauna have been made in many localities. This work, initiated by Sir Wm. Dawson, owes most to W. A. Bell, whose study of Carboniferous flora and fauna spans half a century. Recently, Bell's work has been supplemented by studies of microfaunas (Copeland 1957) and plant spores (Hacquebard 1957; Hacquebard, Barss, and Donaldson 1960).

The problems facing Carboniferous stratigraphers today illustrate the confusing results of attempting to map time-rock rather than rock units.

Bell's subdivision of the Carboniferous rocks of the region has served as the keystone of the stratigraphy on which most subsequent maps have been constructed. His most recent subdivision (Table I) is but a slightly modified version of the one he first proposed in 1925 (Bell 1926).

TABLE I

CARBONIFEROUS STRATA IN THE MARITIME PROVINCES
(after Bell 1958)

Groups	Ages	
Pictou	Westphalian C and D	Pennsylvanian
Local unconformity and disconformity		
Cumberland	Early Westphalian B	
Local unconformity and disconformity		
Riversdale	Westphalian A	
Local unconformity and disconformity		
Canso	Early Namurian and (?) Late Viséan	
Conformity and disconformity		
Windsor	Viséan	Mississippian
Rare local unconformity and disconformity		
Horton	Tournaisian	
Unconformity		

The major units—Horton, Windsor, etc.—were changed in 1944 from a "series" or time-stratigraphic designation to a "group" or rock-stratigraphic designation, in spite of the fact that although some of these units could be defined partly by lithology alone, they were primarily defined by their age as indicated by fossils and were actually time-stratigraphic units as defined by the present-day Code of Stratigraphic Nomenclature. In making the change, Bell stated his philosophy thus (Bell 1944, p. 4): "The economic value of a lithological classification of strata has already been mentioned and is generally understood. The value of a classification based primarily on age is not so generally appreciated. Yet it is the only classification that admits of a logical reconstruction of the geological history of deposition, and that permits of reliable correlation."

Strict adherence to these carefully established time-stratigraphic units in less fossiliferous areas has required the placing of map-unit boundaries where no lithological change is present and has resulted in neglect of the time-transgressive character of many lithological units (Fig. 3). Although

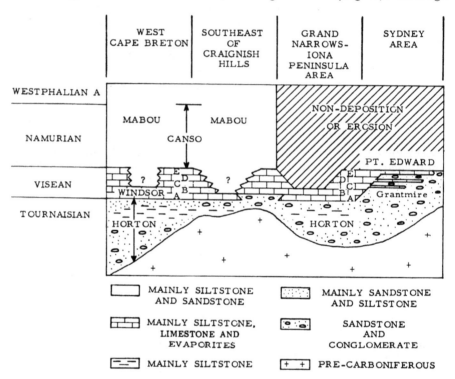

FIGURE 3. Schematic section of lower part of Carboniferous stratigraphy in Cape Breton Island showing time-transgressive nature of rock units.

Bell has long recognized the transgressive nature of lithological units within individual Carboniferous "groups," he has not apparently fully appreciated the transgressive and regressive relations between one "group" and another. The contacts between most, if not all, of the "groups" are conformable at some localities and disconformable or unconformable at others. The age of some conformable contacts probably varies from place to place.

The Horton "Group" and its time equivalents, for example, were defined as continental clastic rocks containing early Mississippian flora and underlying earliest marine strata of Windsor age. In the Sydney area, however, the lowest Windsor unit (Subzone A) is not present, and coarse clastic rocks lie between pre-Carboniferous basement and the late lower Windsor strata.

Because at least some of these clastic rocks are probably of Viséan age, the time-restricted term "Horton Group" could not be applied and another term "Grantmire Member" was coined (Bell and Goranson 1938). These rocks can, however, be traced southwestward about 30 miles along strike into lithologically similar conglomerate and sandstone in the Grand Narrows area, where they are overlain by earliest Windsor (Subzone A) limestone (Fig. 3), and, therefore, they belong to the Horton "Group." It is obvious that Horton-type clastic rocks transgress the Tournaisian–Viséan time-plane and, in the Sydney area, are the same age as part of the type Windsor.

Furthermore, by Bell's time-restricted definition, the lowermost formation of the Horton Group in New Brunswick, the Memramcook, cannot remain part of the "Group" because uppermost Devonian flora have recently been discovered in it (D. C. McGregor, pers. comm., 1962).

The Canso "Group" and its time equivalents were defined as clastic rocks overlying marine upper Windsor strata and, if fossiliferous, containing early Namurian fossils. In the Cumberland Basin, northern Nova Scotia, the clastic rocks that overlie marine earliest upper Windsor strata were included in the Windsor "Group" by Bell (1944, p. 8 and Fig. 11; 1958, p. 34) because he interpreted them as late Viséan in age. An arbitrary contact was drawn between these strata and similar conformably overlying strata of probable early Namurian (Canso) age. Clearly, both units form one lithological unit, as suggested by Gussow (1953, pp. 1750–1) in a nearby area of New Brunswick.

Some of the clastic rocks that overlie marine Windsor strata in western Cape Breton Island (Fig. 3) may have been deposited at the same time as Windsor strata elsewhere on the Island (Kelley 1958, p. 349). On the southeast side of Craignish Hills these clastic rocks (Fig. 3) are apparently structurally conformable with earliest Windsor (Subzone A), although the contact may be a fault. Kelley (1957) mapped these clastic rocks as Canso Group, but he soon realized that this was a poorly chosen term because of the age connotations attached to the term "Canso." Bell (written comm. to Kelley, 1958) considered the same clastic rocks to belong to the Riversdale Group, but he later acknowledged that some Canso strata are also present. Actually, these strata range in age from possible Viséan (Subzone B of the Windsor) to Westphalian A; in Figure 3 they are referred to as Mabou Formation. If Bell's time-restricted "groups" were mapped in this area, then contacts would have to be located by age of contained fossils, and the significance of a thick time-transgressive clastic wedge would probably become obscured.

The time-stratigraphic units which Bell has proposed must be retained. The time-surfaces that they delineate remain the important pages in the history of the Carboniferous. However, for a full appreciation of the tectonic events within that time, the time-transgressive nature of many of the rock units must be recognized. Geologists must not slavishly fit their lithological units between surfaces established solely by the age connotations of fossils.

New rock-stratigraphic terms must be proposed that will permit mapping of lithological units without artificial limitations imposed by time-surfaces.

Preliminary work along these lines has already commenced. Belt (1962) has recently attempted to place some of the Carboniferous rock and time units in their proper perspective. He has raised the status of Mabou Formation (Norman 1935) to Mabou Group, has defined it strictly in lithological terms, and has extended it over all of Nova Scotia. It now includes strata previously assigned to all or part of the Windsor, Canso, and Riversdale "Groups." Belt's major proposals are sound in principle and are the type of study and solution of the problem that we favour. Kelley (1963) has proposed similar revisions for Carboniferous stratigraphy of Cape Breton Island, as outlined in Figure 3.

Isotopic Dating

The initiation of "absolute age" determinations by measurement of isotope ratios of potassium-40–argon-40 (K–Ar) and rubidium-87–strontium-87 (Rb–Sr) was welcomed enthusiastically by workers in the Canadian Appalachians. The possibility of obtaining ages from common minerals such as feldspars and micas and of accurately extrapolating the age of rocks from well-known, fossiliferous terranes into unfossiliferous, and igneous and metamorphic terranes appeared a near panacea to problems of earth history. Eventually these methods or varieties of them may indeed be the major keys to earth history. However, perfection has not yet been achieved, and although isotopic ages have proved to be an invaluable tool, they have also created problems that, for the present, rely on classical geological arguments for their explanation.

Initial Studies

A fortunate impetus was given to geological studies in the Canadian Appalachian region when in the mid-1950's the granitic rocks of Nova Scotia were used by scientists of the Massachusetts Institute of Technology in the pilot studies of the K–Ar and Rb–Sr methods.

Most of the earliest results fell within a 350–400 m.y. range (Fairbairn 1957) suggesting an Ordovician age on the then-accepted time-scale (Holmes 1947). But, some of these granitic rocks were known to cut Lower Devonian strata. A minimum age of Lower Devonian rocks was therefore established and became a major contribution to later revisions of the time-scale (Holmes 1959; Kulp 1961). The bar diagram of M.I.T. results (Fairbairn, Hurley, Pinson, and Cormier 1960, p. 405) on specimens from 27 localities shows 24 within the 300–400 m.y. range with a distinct concentration in the 360–370 m.y. interval. Ages from nearly all specimens collected near the intrusive contacts with Lower Devonian strata fell within this peak interval.

Anomalously low K–Ar and Rb–Sr ages, one of 245 m.y. and several in the 315–340 m.y. range, were obtained from micas in granites of southern-most Nova Scotia. Explanations for these ages will be offered later; suffice it to say here that on stratigraphic evidence the granites are post-Lower Ordovician.

Anomalously high K–Ar and Rb–Sr ages in the 470–518 m.y. range were obtained from micas and feldspar in granites of east-central Cape Breton Island. The granites apparently cut fossiliferous Middle Cambrian strata, which are structurally conformable with overlying lowermost Ordovician strata (Hutchinson 1952; Weeks 1954). Fragments of similar granite occur within the Middle Devonian McAdam Lake Formation. The granite may actually be Ordovician and related to the Taconic orogeny, but further speculation must await closer study. Another possible explanation is proposed later.

In addition to their Nova Scotia studies, M.I.T. scientists determined two Rb–Sr ages of feldspar from Newfoundland granites (Fairbairn 1958). One, from Burin Peninsula, gave a 340 m.y. date that agreed reasonably well with the Nova Scotia pattern; the other, from the Holyrood granite on Avalon Peninsula, gave a 910 m.y. date which substantiated the geological interpretation of a Precambrian age (McCartney 1954) and showed a rather unexpected similarity to the Grenville ages of Quebec and Ontario.

Later M.I.T. studies on the granitic rocks of New Brunswick (Tupper and Hart 1961) provided K–Ar ages in the range 364–398 m.y. These are similar to but a little higher than most ages of the Nova Scotia granites.

Lead isotope age determinations were made at several localities, chiefly in mining camps such as Bathurst, N.B., and Buchans, Nfld., concurrently with the initial M.I.T. isotope age studies in the region (Wilson 1956). Some of these lead ages agree with the M.I.T. K–Ar and Rb–Sr ages for Devonian (Acadian) granites, but some differ greatly and for no reason apparent at present. They are not considered further in this paper.

Subsequent Studies and Resulting Pattern of Ages

Following and partly concurrent with the M.I.T. studies in Nova Scotia, the Geological Survey of Canada began a programme of K–Ar studies in the Appalachian region of Canada. The results of the studies have been reported annually (Lowdon 1960, 1961; Lowdon, Stockwell, Tipper, and Wanless 1963; Leech, Lowdon, Stockwell, and Wanless 1963). M.I.T. scientists have recently applied Rb–Sr whole-rock isochron methods to the Monteregian igneous plutons in southern Quebec (Fairbairn 1962) and to volcanic rocks at the base of the palaeontologically dated Arisaig Group of Nova Scotia (Bottino 1962). Also, University of Alberta scientists have determined K–Ar ages on several minerals from Lower Devonian bentonitic strata of easternmost Gaspé (Smith, Baadsgaard, Folinsbee, and Lipson 1961).

The results of both the initial and subsequent studies are combined in a bar diagram (Fig. 5), and many of the ages are shown on a sketch-map

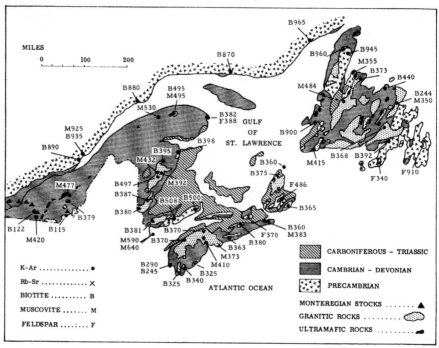

FIGURE 4. Representative K–Ar and Rb–Sr isotopic ages (in millions of years).

of the region (Fig. 4). The bar diagram is similar to that which illustrated the initial M.I.T. age studies (Fairbairn, Hurley, Pinson, and Cormier 1960, Fig. 2). The "peak" for Devonian (Acadian) granites and metamorphic rocks is virtually the same in the two diagrams, although the spread is somewhat greater and the average somewhat older in Figure 5. In addition, there are prominent "peaks" that represent Ordovician (Taconic) and Cretaceous (Monteregian) intrusion and/or metamorphism, and a number of Precambrian ages typical of the Grenville orogenic province (Stockwell 1962). These differences were to be expected, not only because a larger, more geologically diverse area was sampled, but also because particular pains were taken to identify known or suspected pre-Devonian intrusions and metamorphism.

By yielding ages that correspond to Grenville, Taconic, Acadian, and, possibly, Late Palaeozoic orogenic and epeirogenic activity, the K–Ar and Rb–Sr studies have substantiated the tectonic history of the region that was founded on classical geological studies (Weeks 1957). In addition, these isotopic age studies have permitted extrapolation of tectonic episodes into unfossiliferous areas or those known only through gross reconnaissance study (Neale, Béland, Potter, and Poole 1961).

Granites of the Precambrian block which underlie the Great Northern Peninsula of Newfoundland and an inlier of similar rock at St. Georges

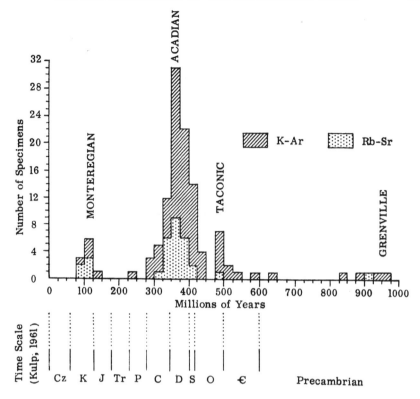

FIGURE 5. Bar diagram of isotopic ages (121 specimens).

Bay have provided K–Ar ages on micas in the range 830–960 m.y. This strengthens the hypothesis that the gneisses, schists, and anorthositic rocks of the Northern Peninsula are an extension of the Grenville Province, as previously suspected on lithological and regional tectonic bases. Based on the Rb–Sr 910 m.y. age on feldspar, a further extension to lithologically different granites and volcanic rocks of southeastern Newfoundland is also suggested.

Several K–Ar ages on micas from igneous and metamorphic rocks of known and probable Precambrian age along the Bay of Fundy fall into the range 500–640 m.y. (Lowdon, Stockwell, Tipper, and Wanless 1963; Leech, Lowdon, Stockwell, and Wanless 1963). Those located northeast of Saint John are relatively low in the range and were obtained from biotites, whereas those located on Grand Manan and nearby islands are relatively high in the range and were obtained from muscovites. Although an Ordovician orogeny cannot be completely ruled out, a possible explanation is that the micas are Precambrian and that some of the radiogenic argon was driven out of the lattice during Devonian orogeny, thus reducing the K–Ar age. The anomalous ages in the 470–518 m.y. range in east-central Cape Breton Island discussed earlier may have a similar explanation.

Probably the most significant Ordovician K–Ar age is the 495 m.y. obtained on both biotite and muscovite in metamorphic rock collected from the aureole of the Mt. Albert ultramafic pluton in Gaspé (Lowdon, Stockwell, Tipper, and Wanless 1963). Ultramafic rocks have been regarded as the earliest intrusion of geosynclinal regions and have generally been considered related to Taconic orogeny throughout the Appalachian system (Hess 1939). That this first quantitative estimate of the age of ultramafic intrusion showed concordant ages for the mineral pair and also coincided with Ordovician time on a recently published time-scale (Kulp 1961) is most encouraging. Subsequent studies of suspected Ordovician granitic bodies that are enclosed within and cut ultramafic rocks in the Eastern Townships of Quebec have given K–Ar ages of 477 and 481 m.y. (Poole, Béland, and Wanless 1963). In central New Brunswick, one sample of old granite has provided a K–Ar age on biotite of 497 m.y. K–Ar ages on micas of other samples of the granite and of associated gneiss fall in the 410–435 range,* indicating the effects of reheating by Devonian granites (Poole 1963; Lowdon, Stockwell, Tipper, and Wanless 1963, pp. 111–12; Leech, Lowdon, Stockwell, and Wanless 1963, pp. 99–102). As all of these Ordovician ages were anticipated on geological evidence and hypotheses, similar ages obtained in less well known areas (e. g. the 484 m.y. age in west-central Newfoundland in the belt from Port aux Basques to the Burlington Peninsula) inspire the further search for geological field evidence of the Taconic orogeny.

The abundance of Devonian K–Ar and Rb–Sr ages obtained on granitic and metamorphic rocks from widely separated localities throughout the region has strengthened the long-prevalent hypothesis that Acadian orogeny was a climactic orogeny that affected almost all parts of the Appalachian region of Canada southeast of the foreland. Although a "peak" age exists between 350 and 375 m.y., very similar to the initial M.I.T. "peak," the broad spread of ages shown in Figure 5 suggests that intrusion may not have been restricted to a single, synchronous pulse. The spread of the peak is small, however, in contrast to the ±30 m.y. analytical error applied to single determinations (Leech, Lowdon, Stockwell, and Wanless 1963, pp. 1–2).

As already noted, most Acadian granites in New Brunswick provide ages on micas in the 380–395 m.y. range, which are from 20 to 40 m.y. older than the 350–370 m.y. range of many Nova Scotia granites. A similar pattern has been recognized in the New England States (Faul, Stern, Thomas, and Elmore 1963). The validity of the pattern must be tested more thoroughly and its significance examined. A recently determined age of 410 m.y. from central Nova Scotia does not fit the pattern. It is too early to conclude that the New Brunswick Acadian granites were intruded

---

*Since this paper was written, a K–Ar age determination of 463 m.y. was obtained from muscovite in a pegmatite associated with the old granite in central New Brunswick, and one of 465 m.y. on biotite from the St. Stephen gabbro body in southernmost New Brunswick (R. K. Wanless, pers. comm., 1963).

and cooled 20 to 40 m.y. earlier than those of Nova Scotia. The relatively unknown effects of the duration of cooling of intruded masses and of epeirogenic uplift and downwarp (Hurley, Hughes, Pinson, and Fairbairn 1962) on K–Ar age determinations may have contributed to the present pattern.

Ages suggestive of Late Palaeozoic intrusive and/or metamorphic activity have been obtained in Newfoundland and Nova Scotia. In east-central Newfoundland, one sample yielded ages of 244 m.y. on highly chloritized biotite and, subsequently, 350 m.y. on muscovite. The 245 m.y. K–Ar age from granite in southernmost Nova Scotia was also on chloritized biotite (Fairbairn, Hurley, Pinson, and Cormier 1960, p. 405). As a check, the age of biotite in another specimen from the same pluton was later determined by K–Ar as 290 m.y. (Lowdon, Stockwell, Tipper, and Wanless 1963, p. 114). Other K–Ar and some Rb–Sr determinations on micas from granites in southernmost Nova Scotia fall in the range of 315–340 m.y., significantly lower than the 360–410 m.y. range of ages in central and eastern Nova Scotia. The granites in southernmost Nova Scotia occur within strata regionally metamorphosed to a much higher grade than is typical elsewhere in the Meguma Group. Perhaps southernmost Nova Scotia was affected by an otherwise undetected Late Palaeozoic metamorphism, during which the micas became partly or completely recrystallized and degassed, i.e. radiogenic argon was driven out. No evidence is available in the vicinity to bear on this hypothesis. However, as noted earlier, the late Palaeozoic intrusion in Rhode Island, U.S.A., the sills and plugs which cut Upper Mississippian Mispek Group near Saint John, N.B., and the Belleoram granite* which cuts Upper Devonian conglomerates near Fortune Bay, southeastern Newfoundland, all lend support to the possibility of a zone of Late Palaeozoic metamorphism passing through southernmost Nova Scotia.

The ultra-alkaline plugs which form the Monteregian Hills of southern Quebec have been investigated using both K–Ar and Rb–Sr methods. On geological grounds they are known to be post-Devonian. The average of five K–Ar age determinations is 114 m.y.; the average of 10 Rb–Sr ages on biotite is 106 m.y., and the Rb–Sr whole-rock isochron age is 116 m.y. These ages are Cretaceous on all the more recent time-scales (Fairbairn 1962).

Problems Posed by Some Isotopic Ages

Since the beginning of the isotopic age studies, anomalously low and high ages have been obtained for which no explanation was apparent in terms of the geological setting. Many of these ages fell within the generous margin of analytical error allowed by the physicists for single determinations; for Ordovician and Devonian ages, the error is $+35$ to $+30$ m.y.

*A recently determined K–Ar age of 400 m.y. was obtained from chloritized biotite from Belleoram granite (R. K. Wanless and F. D. Anderson, pers. comm., 1963).

(Leech, Lowdon, Stockwell, and Wanless 1963, pp. 1–2) or about the order of magnitude of one period in the Middle Palaeozoic. The possibility of even larger analytical errors is seldom seriously considered—although, in fact, it is conceivable. Some apparently anomalous ages can, of course, be reconciled by revision of the time-scale to suit the ages and field relations.

Anomalously low K–Ar and Rb–Sr ages on micas, as in southernmost Nova Scotia, can best be explained by suggesting that an orogenic or thermal event has caused heating of the micas. This results in driving the radiogenic daughter elements out of the rock to some unknown and always unspecified place. Hurley, Hughes, Pinson, and Fairbairn (1962) recently suggested that radiogenic argon was not retained in micas held some 9000 feet below the surface prior to epeirogenic uplift along the Alpine Fault in New Zealand. Almost all anomalous ages in the Appalachian region of Canada are younger than expected from field relations and, hence, loss of the daughter elements, particularly argon, is almost an unchallengeable explanation. Thus, all ages are regarded as minimum ages and the interpretation, within the limits of analytical error, becomes a question of how close the "minimum" age approaches the "maximum" or true age. The Rb–Sr whole-rock isochron method now being tested thoroughly in the field shows some promise of overcoming the effects of limited diffusion of parent and daughter elements.

Anomalously high K–Ar and Rb–Sr ages on micas and Rb–Sr ages on feldspars and whole-rock samples can be explained either by loss of parent elements or gain of daughter elements by diffusion long after formation of the rocks, or by absorption of daughter elements from old source materials at the time of formation of the rock. Radiogenic argon has not yet been shown in the Appalachian region of Canada to have been absorbed from source materials or to have been absorbed from an external source after formation of the rock. Limited experience with Rb–Sr ages in the region indicates a parallelism of Rb–Sr and K–Ar ages, thus leading to the conclusion that absorption of relict radiogenic strontium from source materials occurs below detectable levels, if at all. Examples of some of the problematic ages are given below.

A K–Ar age of 530 m.y. has been obtained on muscovite from a phyllite of the Shickshock Group of Gaspé Peninsula (Lowdon, Stockwell, Tipper, and Wanless 1963, p. 106). The group is Lower Ordovician or older (N.C. Ollerenshaw, pers. comm., 1962) and is overlain unconformably by gently dipping Silurian strata. Metamorphism of the group was presumed to be Ordovician (Taconic), but the 530 m.y. age coincides with the Middle–Late Cambrian boundary on a recent time-scale (Kulp 1961). However, application of the 35 m.y. limit of analytical error brings this date to the 495 m.y. muscovite and biotite ages for the Taconic Mt. Albert ultramafic pluton on strike to the east. Furthermore, granitic bodies in some of the ultramafic rocks of southern Quebec which intrude late Middle Ordovician strata have produced K–Ar muscovite ages of 477 and 481 m.y. For this reason, a revision of the time-scale has been proposed (Poole, Béland, and Wanless 1963) which would place the Cambrian–Ordovician boundary at about 535 m.y. With this explanation the 530 m.y. Shickshock date becomes more plausible.

In central New Brunswick, a field hypothesis that a certain granite and related gneiss are older than Devonian (Acadian) granites was tested with several K–Ar ages on micas (Poole 1963; Lowdon, Stockwell, Tipper, and Wanless 1963, pp. 111–12; Leech, Lowdon, Stockwell, and Wanless 1963, pp. 99–102). Ages ranged from 410 to 435 m.y. on five samples located from 0.4 to 3.3 miles from the contacts with typical Devonian granites. A sixth and very different age of 497 m.y. was obtained from biotite, only 0.5 mile from the same contact. Ages on muscovite were from 6 to 25 m.y. older than ages on nearby biotite. There is no obvious regular decrease of age (loss of argon) for those samples progressively closer to the contact. Perhaps the 497 m.y. age is grossly in error.

Elsewhere in the Appalachian region of Canada, an increasing number of determined ages fall in the 400–440 m.y. range, and their interpretation and significance is a vital problem. Ages near either end of the range can be interpreted as either Ordovician (Taconic) or Devonian (Acadian) within the limits of analytical error. Indeed, the limits of $+30$ m.y. span the interval, and raise the question of whether single determinations by the K–Ar method are accurate or reliable enough to render a sufficient degree of confidence in attempting to distinguish between early Devonian, Silurian, and late Ordovician events. However, to a degree, the method can distinguish between these ages, because the limits of analytical error are commonly much less than $+30$ m.y. wherever either duplicate samples or comparable samples of the same rock nearby have been analysed.

Ages in the 400–440 m.y. range continue to emphasize the possibility that some intrusion and metamorphism may be Silurian, and neither Devonian nor Ordovician as is generally assumed. For example, a 440 m.y. K–Ar age on biotite from granodiorite which cuts the fossiliferous Silurian Botwood Group (Leech, Lowdon, Stockwell, and Wanless 1963, pp. 119–20) in northern Newfoundland falls beyond the limits of analytical error expected for a Devonian granite.

The geological gymnastics required to interpret some isotopic ages were demonstrated by Neale and Nash (1963). The granitic rocks of the Sandy Lake region of west-central Newfoundland have for many years been interpreted as Devonian. Of four K–Ar ages determined, three fell in the range 355–373 m.y. and are Devonian, and one was 484 m.y., and is Ordovician. Subsequent to the age determinations, re-examination of the field data indicated that the altered granodiorite, which yielded the 373 m.y. age, cuts Ordovician rocks and is unconformably overlain by deformed Silurian strata. The anomalous 373 m.y. age was explained as a result of updating (i.e. degassing) during the superimposed Devonian intrusion and metamorphism. Conversely, the granite which yielded the 484 m.y. age is not different lithologically and structurally from typical Devonian (Acadian) granite, although Neale and Nash tentatively interpreted it as Ordovician solely on its isotopic age!

The Rb–Sr whole-rock isochron method appears to be the most satisfactory method of isotopic age study developed to date (Compston and Jeffery 1959; Fairbairn, Hurley, and Pinson 1961). It seems to be unaffected by millimetric diffusion of parent and daughter isotopes and will "see through" or "penetrate" the effects of weak metamorphism and thus provide true ages for major tectonic events rather than the "minimum" ages given by the K–Ar and Rb–Sr single-mineral method. However, the pioneer whole-rock isochron study by M.I.T. scientists in the Appalachian region of Canada has already raised major problems. A preliminary Rb–Sr whole-rock isochron age of the Arisaig volcanic rocks is 400 m.y. (Bottino 1962). Stratigraphically, these rocks have a minimum age of Early Silurian —i.e. about 425 m.y. on the present time-scale. Moreover, an isochron age on late Lower Devonian rhyolite from Maine places the base of the Devonian at about 365 m.y. Changes in the time-scale or in the constants used in calculating the ages are suggested, but it is too early in the study to appraise these results fully.

Age and Correlation by Palaeomagnetism

Studies of palaeomagnetism in the Appalachian region of Canada have been carried out by P. M. Dubois, A. Larochelle, and R. F. Black of the Geological Survey of Canada, and a study was undertaken in Newfoundland by Nairn, Frost, and Light (1959). The Geological Survey studies have been directed mainly towards establishing a reliable curve of polar wandering for North America, as described in another paper of this symposium (Morley and Larochelle, pp. 39–51), but, they have also shed light on both age and correlation problems and clarified a problem in geotectonics.

Successful age studies were conducted by Larochelle (1961, 1962) on the Monteregian intrusions of southern Quebec. He was able to show, on the basis of palaeo-pole positions, that the collar of more basic rocks on Mount Megantic was actually of the same age as the Monteregian suite. He also showed that pole positions of these rocks corresponded closely to post-Triassic poles from elsewhere in North America and fell nearest to the established Cretaceous pole position. This conclusion was corroborated independently by the isotopic age determinations shown on Figures 4 and 5.

A problem of correlation of red beds of the Silurian (?) Springdale Group in north-central Newfoundland has been solved by Black, who will publish the data in the near future. Isolated outcrop areas of red beds in the Green Bay – Halls Bay region have been correlated with the Springdale Group (MacLean 1947), but recently it was suggested (Neale, Nash, and Innes 1960) that the red beds near King's Point village were much less indurated than typical Springdale beds and therefore probably much younger. Black (pers. comm., 1963) showed that palaeo-poles from King's Point strata have the same general orientation as poles from the Springdale Group and differ only by an inversion of polarity, which has no important time-significance. By subsequent petrographic examination of these rocks, differences in induration were explained as local variations in cement. Comparative provenance studies and other geological data have substantiated Black's conclusions (Neale and Nash 1963).

A larger problem was brought forward by Nairn, Frost, and Light (1959) when, after their determination of palaeomagnetic pole positions for certain Precambrian and Carboniferous rocks of Newfoundland, they suggested that Newfoundland may have been rotated 20 degrees anticlockwise relative to the rest of North America in post-Carboniferous time. Their tentative conclusion was refuted by DuBois (1959), whose studies of Carboniferous rocks from New Brunswick and Gaspé Peninsula showed that there was essentially no difference in palaeomagnetic pole positions from those of Newfoundland and, hence, no evidence of post-Carboniferous rotation. DuBois' conclusions are in agreement with geological evidence (King 1951, pp. 90–91), which suggests the possibility of rotation in pre-Carboniferous time but of no rotation in Carboniferous and post-Carboniferous time.

CONCLUSIONS

In general, in the Appalachian region of Canada good agreement exists between the classical methods of dating and correlation and the newer methods based on decay of unstable isotopes and the shift of magnetic pole positions throughout geological time.

Few fossil localities have been recorded in the Lower Palaeozoic rocks of most of the region and much reliance is still placed on lithological, structural, and igneous correlations. It is a tribute to the unifying concepts of geology that such correlations are so often substantiated when palaeontological and isotopic dates are finally obtained.

"To conclude, Appalachian geology, although of advancing years, is in robust health with fond memories of an interesting and eventful past but with a sharp eye to the future and new fashions . . ."—Contributed by Professor D. M. Baird, F.R.S.C., who ably read this paper to the Society in June 1963.

REFERENCES

ALCOCK, F. J. (1938). Geology of Saint John Region, New Brunswick. Geol. Surv. Can., Mem. 216.
—— — (1941). Jacquet River and Tetagouche River map-areas, New Brunswick. Geol. Surv. Can., Mem. 227.
—— — (1947). The Appalachian Region. Geol. Surv. Can., Econ. Geol. Ser., no. 1, pp. 98–155.
——— (1960). St. Stephen, Charlotte County, New Brunswick. Geol. Surv. Can., Map 1096A.

AMERICAN COMMISSION ON STRATIGRAPHIC NOMENCLATURE (1961). Code of strati-graphic nomenclature. Am. Assoc. Petrol. Geol. Bull., *45*, no. 5.

AMI, H. M. (1903). On the Upper Cambrian age of the *Dictyonema* slates of Angus Brook, New Canaan and Kentville, Nova Scotia. Proc. and Trans., N.S. Inst. Sci., *10*: 447–50.

ANDERSON, F. D. and POOLE, W. H. (1959). Geology of the Woodstock-Fredericton area, New Brunswick. Geol. Surv. Can., Map 37–1959.

BÉLAND, J. (1960). Rimouski–Matapédia area. Quebec Dept. Mines, Prelim. Rept. 430.

BELL, W. A. (1926). Carboniferous formations of Northumberland Strait, Nova Scotia. Geol. Surv. Can., Sum. Rept., pt. C, pp. 142–79.

———— (1944). Carboniferous rocks and fossil floras of northern Nova Scotia. Geol. Surv. Can., Mem. 238.

———— (1958). Possibilities for occurrence of petroleum reservoirs in Nova Scotia. Nova Scotia Dept. of Mines.

BELL, W. A. and GORANSON, E. A. (1938). Sydney Sheet, west half, Nova Scotia. Geol. Surv. Can., Map 360A.

BELT, E. (1962). Stratigraphy and sedimentology of the Mabou Group (Middle Car-boniferous), Nova Scotia. Ph.D. thesis, Yale Univ., unpubl.

BERRY, W. B. N. (1962). On the Magog, Quebec, graptolites. Am. J. Sci., *260*: 142–8.

BILLINGS, M. P. (1956). The Geology of New Hampshire, pt. 2. Bedrock Geology. N.H. State Planning Comm., Concord, N.H.

BOTTINO, M. L. (1962). Whole-rock Rb–Sr studies on volcanics and some related granites. Mass. Inst. Technol., NYO-3943, 10th Ann. Progr. Rept., 1962, U.S. Atomic Energy Comm. Cont. AT(30–1)–1381, pp. 51–4.

BOUCOT, A. J. and CUMMING, L. M. (1963). Contributions to the age of the Gaspé sandstone. Geol. Surv. Can., Bull. (in press).

COMPSTON, W. and JEFFERY, P. M. (1959). Anomalous common strontium in granite. Nature, *184*: 1792.

COOPER, J. R. (1954). La Poile – Cinq Cerf map-area, Newfoundland. Geol. Surv. Can., Mem. 276.

COPELAND, M. J. (1957). The Arthropod fauna of the Upper Carboniferous rocks of the Maritime Provinces. Geol. Surv. Can., Mem. 286.

CROSBY, D. G. (1952). Preliminary map Wolfville (east half) Hants and Kings Counties, Nova Scotia. Geol. Surv. Can., Paper 52–18.

———— (1963). Wolfville map-area, Nova Scotia. Geol. Surv. Can., Mem. 325.

DuBois, P. M. (1959). Palaeomagnetism and rotation of Newfoundland. Nature, *184*: 63–4.

DUQUETTE, G. (1960). Gould area, Quebec. Quebec Dept. Mines, Prelim. Rept. 432.

FAIRBAIRN, H. W. (1957). Nova Scotia mica ages. Mass. Inst. Technol., 4th Ann. Progr. Rept., 1956–57, U.S. Atomic Energy Comm., Cont. AT(30-1)-1381, pp. 24–8.

———— (1958). Age data from Newfoundland. Mass. Inst. Technol., 5th Ann. Progr. Rept., 1957–58, U.S. Atomic Energy Comm., Cont. AT (30–1)-1381, p. 69.

———— (1962). Past and present magnitude of Sr^{87}/Sr^{86} in rocks and minerals of the Monteregian Igneous Province, Quebec. Mass. Inst. Technol., NYO-3943 10th Ann. Progr. Rept., 1962, U.S. Atomic Energy Comm., Cont. AT(30-1)-1381, pp. 55–60.

FAIRBAIRN, H. W., HURLEY, P. M., and PINSON, W. H. (1961). The relation of dis-cordant Rb–Sr mineral and whole-rock ages in an igneous rock to its time of crystallization and to the time of subsequent Sr^{87}/Sr^{86} metamorphism. Geochim. Cosmochim. Acta, *23*: 135–44.

FAIRBAIRN, H. W., HURLEY, P. M., PINSON, W. H., and CORMIER, R. F. (1960). Age of the granitic rocks of Nova Scotia. Geol. Soc. Am. Bull., *71*: 399–414.

FAUL, H., STERN, T. W.. THOMAS, H. H., and ELMORE, P. L. D. (1963). Ages of intrusion and metamorphism in the northern Appalachians. Am. J. Sci., *261*: 1–19.

GREINER, H. R. (1960). Pointe Verte, Gloucester and Restigouche counties, New Brunswick. N.B. Dept. Lands and Mines, P.M. 60–2.

GUSSOW, W. C. (1953). Carboniferous stratigraphy and structural geology of New Brunswick, Canada. Bull. Am. Assoc. Petrol. Geol., *37*: 1713–816.

HACQUEBARD, P. A. (1957). Plant spores in coal from the Horton Group (Mississippian) of Nova Scotia. Micropalaeontology, *3* (4): 301–24.

HACQUEBARD, P. A., BARSS, M. S., and DONALDSON, J. R. (1960). Distribution and stratigraphic significance of small spore genera in the Upper Carboniferous of the Maritime Provinces of Canada. Compt. Rend., 4th Intern. Congr. on Carb. Strat. and Geol., Heerlen, T.I., 237–45.

HESS, H. H. (1939). Island arcs, gravity anomalies and serpentinite intrusions. Intern. Geol. Congr. Moscow, 1937, Rept. 17, vol. 2, pp. 263–83.

HOLMES, A.(1947). The age of the Earth. Endeavour, *6* (23): 99–108.

——— (1959). A revised geological time-scale. Trans. Edin. Geol. Soc., *17* (pt. 3): 183–216.

HURLEY, P. M., FAIRBAIRN, H. W., PINSON, W. H., and FAURE, G. (1959). K–A and Rb–Sr minimum ages for the Pennsylvanian section in the Narragansett Basin. Mass. Inst. Technol. 7th Ann. Rept., 1959, U.S. Atomic Energy Comm., Cont. AT(30-1)-1381, pp. 97–121.

HURLEY, P. M., HUGHES, H., PINSON, W. H., and FAIRBAIRN, H. W. (1962). Radiogenic argon and strontium diffusion parameters in biotite at low temperatures obtained from Alpine fault uplift in New Zealand. Geochim Cosmochim. Acta, *26*: 67–80.

HUTCHINSON, R. D. (1952). The stratigraphy and trilobite faunas of the Cambrian sedimentary rocks of Cape Breton Island, Nova Scotia. Geol. Surv. Can., Mem. 263.

——— (1963). Cambrian stratigraphy and trilobite faunas of Newfoundland. Geol. Surv. Can., Bull. 88.

KELLEY, D. G. (1957). Whycocomagh, Nova Scotia. Geol. Surv. Can., Map 17–1957.

——— (1958). Mississippian stratigraphy and petroleum possibilities of central Cape Breton Island, Nova Scotia. Can. Inst. Mining Met. Bull., *51*: 341–51.

——— (1963). Baddeck and Whycocomagh map-areas with emphasis on Mississippian stratigraphy of central Cape Breton Island, Nova Scotia. Geol. Surv. Can., Mem. (in press).

KING, P. B. (1951). The tectonics of middle North America. Princeton: Princeton University Press.

KULP, J. L. (1961). Geologic time-scale. Science, *133*: 1105–14.

LAROCHELLE, A. (1961). Application of palaeomagnetism to geological correlation. Nature, *192*: 37–9.

——— (1962). Palaeomagnetism of the Monteregian Hills, southeastern Quebec. Geol. Surv. Can., Bull. 79.

LEECH, G. B., LOWDON, J. A., STOCKWELL, C. H., and WANLESS, R. K. (1963). Age determinations and geological studies, Rept. 4. Geol. Surv. Can., Paper 63-17.

LOWDON, J. A. (1960). Age determinations by the Geological Survey of Canada, Rept. 1. Geol. Surv. Can., Paper 60-17.

——— (1961). Age determinations by Geological Survey of Canada, Rept. 2. Geol. Surv. Can., Paper 61-17.

LOWDON, J. A., STOCKWELL, C. H., TIPPER, H. W., and WANLESS, R. K. (1963). Age determinations and geological studies, Rept. 3. Geol. Surv. Can., Paper 62-17.

MACLEAN, H. J. (1947). Geology and mineral deposits of the Little Bay area, St. John's. Nfld. Geol. Surv., Bull. 22.

MARLEAU, R. A. (1959). Age relations in the Lake Megantic range, southern Quebec. Proc. Geol. Assoc. Can., *11*: 129–39.

McCARTNEY, W. D. (1954). Holyrood, Newfoundland. Geol. Surv. Can., Paper 54–3.

McGERRIGLE, H. W. (1954). The Tourelle and Courcelette areas, Gaspé Peninsula. Quebec Dept. of Mines, Geol. Rept. 62.

McLEARN, F. H. (1924). Palaeontology of the Silurian rocks of Arisaig, Nova Scotia. Geol. Surv. Can., Mem. 137.

NAIRN, A. E. M., FROST, D. V., and LIGHT, B. G. (1959). Palaeomagnetism of certain rocks from Newfoundland. Nature, *183*: 596–7.

NEALE, E. R. W. (1957). Ambiguous intrusive relationship of the Betts Cove – Tilt Cove serpentine belt, Newfoundland. Proc. Geol. Assoc. Can., *9*: 95–107.

NEALE, E. R. W., NASH, W. A., and INNES, G. M. (1960). King's Point, Newfoundland. Geol. Surv. Can., Map 60-35.

NEALE, E. R. W., BÉLAND, J., POTTER, R. R., and POOLE, W.H. (1961). A preliminary tectonic map of the Canadian Appalachian region based on age of folding. Can. Inst. Mining Met. Bull., *54*: 687–94.

NEALE, E. R. W. and NASH, W. A. (1963). Sandy Lake (east half), Newfoundland. Geol. Surv. Can., Paper 62-28.

NORMAN, G. W. H. (1935). Lake Ainslie map-area, Nova Scotia. Geol. Surv. Can., Mem. 177.

NORTHROP, S. A. (1939). Paleontology and stratigraphy of the Silurian rocks of the Port Daniel – Black Cape Region, Gaspé. Geol. Soc. Am. Bull., Sp. Paper no. 21.

POOLE, W. H. (1960). Hayesville and McNamee map-areas, York, Northumberland, and Carleton counties, New Brunswick, 21J/10 and 21J/9 (W 1/2). Geol. Surv. Can., Paper 60-15.

———— (1963). Hayesville, New Brunswick. Geol. Surv. Can., Map 6-1963.

POOLE, W. H., BÉLAND, J., and WANLESS, R. K. (1963). Minimum age of Middle Ordovician rocks in southern Quebec. Geol. Soc. Am. Bull., *74*: 1063–6.

SMITH, C. H. (1957). Bathurst-Newcastle area, N.B. Geol. Surv. Can., Map 1-1957.

SMITH, D. G. W., BAADSGAARD, H., FOLINSBEE, R. E., and LIPSON, J. (1961). K/Ar age of Lower Devonian bentonites of Gaspé, Quebec, Canada. Geol. Soc. Am. Bull., *72*: 171–4.

SMITHERINGALE, W. (1960). Geology of Nictaux–Torbrook map-area, Annapolis and Kings counties, Nova Scotia. Geol. Surv. Can., Paper 60-13.

STOCKWELL, C. H. (1962). A tectonic map of the Canadian Shield. *In* The Tectonics of the Canadian Shield, *edited by* J. S. Stevenson (Roy. Soc. Can. Spec. Pub. No. 4), pp. 4–15. Toronto: University of Toronto Press.

TAYLOR, F. C. (1962). Annapolis, Nova Scotia. Geol. Surv. Can., Map 40-1961.

TUPPER, W. M. and HART, S. R. (1961). Minimum age of the Middle Silurian in New Brunswick based on K–Ar method. Geol. Soc. Am. Bull., *72*: 1285–8.

TWENHOFEL, W. H. and SHROCK, R. R. (1937). Silurian strata of Notre Dame Bay and Exploits Valley, Newfoundland. Geol. Soc. Am. Bull., *48*: 1743–84.

WEEKS, L. J. (1954). Southeast Cape Breton Island, Nova Scotia. Geol. Surv. Can., Mem. 277.

———— (1957). Appalachian Region; Geology and economic minerals of Canada, 4th ed. Geol. Surv. Can., Econ. Geol. Ser. no. 1, pp. 123–205.

WILLIAMS, H. (1962). Botwood (west half) map-area, Newfoundland. Geol. Surv. Can., Paper 62-9.

———— (1963). Twillingate map-area, Newfoundland. Geol. Surv. Can., Paper 63-21.

WILSON, J. T. (1956). Economic significance of basement subdivision and structures in Canada. Can. Inst. Mining Met. Bull. no. 532, pp. 550–8.

WOODMAN, J. E. (1908). Probable age of the Meguma (gold-bearing) series of Nova Scotia. Geol. Soc. Am. Bull., *19*: 99–112.

HISTORICAL GEOLOGY OF THE DEVONIAN SYSTEM IN WESTERN CANADA

R. de Wit

ABSTRACT

The elements of the stratigraphic succession vary from place to place. They must be established locally until palaeontological, tectonic, and sedimentological data permit a comparison of different sections. The chronological order of major episodes in geological history may, thus, be established. The study of Devonian sediments in western Canada may be cited as an example. The observations recorded by early geological explorers have been augmented by a vast number of newly discovered data, and a certain pattern is now perceived of Devonian geological events and their relative significance in time and space.

DETAILED PALAEONTOLOGICAL STUDIES and absolute age determinations offer undoubtedly the most direct approach to geochronology. Where these methods fail, however, the stratigrapher is forced to determine a chronological order based on stratigraphy alone. This can be done through study of the geological record of regional earth movements. The procedure that is followed may be compared with the mathematical problem of solving n equations with n unknowns, the time intervals being the unknowns while the given data include all lithological, biological, and tectonic features or, in short, the facies of sedimentary rocks.

As is usual in geology, the unknown factors exceed by far the number of observed or measured data. Although our method promises, therefore, only limited success, we must investigate its possibilities.

The first step in stratigraphic studies is the subdivision of the succession of beds within a certain area into promising lithological units. When the standard section is compared with other sequences it becomes evident that some of the chosen units lose their identity, but other units are wide-spread and may be recognized regardless of considerable changes in composition. Lithological or palaeontological markers are established, and some of these approach the fixation of a moment in geological time, while others are more obviously diachronous. After the distribution of major units is established, one may analyse the significance of their vertical order of occurrence. If it can be proved that the succession of beds corresponds to a regular rhythm in crustal movements, then the stratigraphic analysis may result in the discovery of the pendulum of geologic time.

Devonian Sedimentation

The Devonian system in western and northern Canada contains beds and units of as varied lithologies as one might expect, and Devonian strata have been studied over a wide area by all conceivable methods. It may be justified, therefore, to take stock of lithological and palaeontological data and to review briefly their meaning with regard to the geological history of a part of the earth's crust.

Beds of Devonian age in western Canada have been studied since about 1873, when geologists of the Geological Survey of Canada began their famous exploits. Their successors have vigorously maintained this tradition of geological pioneering. Others from the academic world contributed considerably to the Devonian studies but, after oil was discovered in 1947 in the Devonian of central Alberta, a significant increase in research and knowledge resulted from the joint efforts of all students of sedimentary rocks in western Canada.

Before an analysis of the historical geology of the Devonian of western Canada is made, we must briefly review the raw material at hand.

South of Great Slave Lake and the Mackenzie Mountains, Devonian beds consist mainly of carbonate rocks, calcareous shales, salts, and silts, but farther to the north the products of evaporation are less conspicuous and, here, a substantial part of the section is composed of continental and marine quartz sandstones and greywackes. The general distribution of these sediments should be considered in the light of the palaeogeographic setting in which they occur (**Fig. 1**).

The pre-Devonian topography of the continental margin was characterized by plateaux and by the presence of high but gentle ridges and mounds. The Peace River Arch and the Providence "high" are well known, and another ridge may have extended in southwestern Alberta, parallel to the present Rocky Mountain front. These ridges divided the inundated platform into two major provinces, which could be called the northern "open" shelf and the southern "protected" shelf. The dividing line may tentatively be drawn west of Great Slave Lake.

In most areas the stratigraphic sections reveal that the Devonian sea invaded the epicontinental landmass step by step. Devonian sediments cover a wide variety of Palaeozoic beds and Precambrian rocks, which they overlie in most places with pronounced unconformity. However, in the Northwest Territories, the Yukon, and in the Canadian Arctic Archipelago, a few localities are known where the Silurian and Devonian systems probably form a continuous succession. It is believed that these localities were situated near the edge of the continental platform. Here, beds of Lower Devonian age have been observed but our knowledge of their lithology and palaeontology leaves much to be desired, and the following discussion concerns itself mainly with the widespread Middle and Upper Devonian rocks.

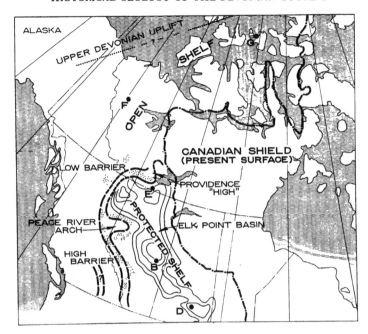

FIGURE 1. Framework of sedimentation on the Devonian shelves of western Canada. B, D, E, F, and G are the approximate locations of typical sections shown in Figures 2 and 3.

Carbonate deposits form an important part of the Devonian sediments. Several different types can be distinguished and each tells a story about its peculiar environment of deposition. Most impressive are the Upper Devonian reef complexes, which form large masses composed of clastic carbonate rocks that have been held in place by buttresses and mats of reef-building organisms. They have relatively steep edges and reach a thickness of more than 1000 feet, indicating that they grew steadily while their flat tops remained close to sea level. Some reef complexes were isolated, such as the Leduc reef, while others formed wide marginal rims around the basins. These marginal reef complexes contain many scattered lenses and impregnations of salt and anhydrite which formed in shallow lagoons. The carbonates in the marginal reefs are generally less fossiliferous than those in the isolated reef complexes.

Other common carbonate units include flat lenses of clastic limestone and dolomite which vary in extent from a few hundred feet to more than 50 miles. Some of these show regular grading from basal fossiliferous beds and flat reefs that were formed in relatively quiet water, to clastic carbonates at the top, deposited under turbulent conditions. These units are considered to be carbonate banks which accumulated during relatively short periods on any favourable shoal, while sea level changed. Small carbonate banks can be examined in exposures along the Athabasca River near McMurray.

They seem to form a continuous layer of fossiliferous limestone with a thickness of 10 to 15 feet. Closer examinations reveal, however, that this band is composed of partly overlapping lenses which fade into shale along their thin edges.

It is interesting to note here that R. G. McConnell observed in 1888 and 1889 the limestone escarpments and high waterfalls that occur on many southern tributaries of the Mackenzie River between Hay River and Liard River. On the basis of close lithological similarity of the outcrops and of their faunal assemblages, he concluded that they were all part of a continuous limestone unit. He included the limestone exposed at the Ramparts near Fort Good Hope in the same unit, although he acknowledged the opinion of his colleague, Whiteaves, that this limestone was possibly older than that exposed in the Hay River – Liard area.

It is now known that the limestones at the Ramparts are Middle Devonian in age while those exposed in the southern Territories are Upper Devonian. Moreover, by surface and subsurface studies it has been demonstrated that these Upper Devonian limestones do not form one continuous unit, but that they represent a series of carbonate banks that are spread like a deck of cards, with the upper ones in the west.

The error made by McConnell is still commonly repeated today, and disputes arise between stratigraphers and palaeontologists about the correlation of some physically identical limestone sections. Later it will be pointed out in which parts of the section these units may most easily be confused. The study of the occurrence, shape, fabric, and components of carbonate banks in western Canada is still in its infancy and should be expanded greatly.

It was mentioned that evaporite deposits are associated with carbonates in the marginal reef complexes. The most prominent bodies of salt and anhydrite are in the eastern part of the "protected" shelf area of Alberta and Saskatchewan, where they can be followed in the subsurface for hundreds of miles. Some are more than 600 feet thick. These evaporite deposits must have formed in isolated pans during slow transgression, as well as during withdrawal of the sea, until most irregularities of the shelf floor were filled up.

In many places a close relationship exists between precipitated salts and the occurrence of silty red beds and siltstones, although the latter do not necessarily contain salt or anhydrite. The silts are well sorted and can be regarded as primarily of aeolian origin. In some localities the red beds represent old soil zones, but in other areas the silts are concentrated in marine deposits that contain limestone or marine shale. These units are generally thin but in places they attain a thickness of several hundred feet. They provide persistent and wide-spread geological markers.

Recently, a detailed study was made of such a zone by D. J. McLaren and E. W. Mountjoy (1962) near Jasper in the Rocky Mountains. Their observations established the fact that a prominent silty formation (formerly mapped as the Alexo) consists of two silty units that are in disconformable contact. The stratigraphic gap involved marks the end of the Frasnian stage and the beginning of the Famennian. Several authors have noticed that the most drastic faunal change in the Devonian system occurs at this level in western Canada. The silty zone may represent, consequently, a long time-interval. A similar conclusion can be reached on purely stratigraphic grounds and the inference seems justified that all Devonian silty red beds of this type were formed slowly during a considerable period of geologic time.

Clastic sediments of terrigenous origin form a major constituent of the Devonian system. The Middle Devonian of the southern protected shelf area, however, contains relatively little shale and sandstone. A thin discontinuous veneer of quartz sands accompanies the ancestral topographic barriers west and north of the basin, where it marks the base of the reduced Middle Devonian or, in its absence, the early Upper Devonian.

In the Northwest Territories and the Arctic Archipelago, shales form a notable proportion of Middle Devonian beds, although the majority of these sediments are carbonate rocks. It is also significant that in the Bird Fiord formation of Givetian age, sands and sandstones occur, which reflect uplift in the western and northwestern Arctic Islands region.

The Upper Devonian rock sequences of the protected shelf, as well as those of the open shelf, contain major terrigenous deposits. In Alberta the Beaverhill Lake and Woodbend intervals contain in some areas 1000 to 2500 feet of shale. Upper Devonian greywackes, quartzitic sandstones, and shales, largely continental deposits, predominate in the Northwest Territories and in part of the Arctic Islands. They form the "molasse" deposits on the flank of a pre-orogenic uplift, and they resemble the Cretaceous clastics that occur along the Rocky Mountain front in Alberta. The greatest observed thicknesses, of the order of 10,000 feet, occur in the northwestern Mackenzie Basin and on Melville and Ellesmere islands.

The continental sequence is interrupted by a few marine tongues. Few fossils have been found in these beds and only preliminary identifications are available. It may be expected that the marine tongues of the northern clastic series correspond to the maximum expansion of the Upper Devonian seas in the regions farther to the south.

Sedimentation and Epeirogenesis

One specific example of a Devonian succession of beds will now be examined. A generalized composite section which can be considered as representative for central or east-central Alberta is illustrated in Figure 2A.

It is not difficult to discern in the sedimentary sequence a basic cyclic pattern of transgression and regression. In the Elk Point group, which forms the lower part of the succession, two main carbonate groups represent "peaks" of transgression. The salt deposits were formed in the initial and late phases of transgression. Probably the red beds and silts were deposited

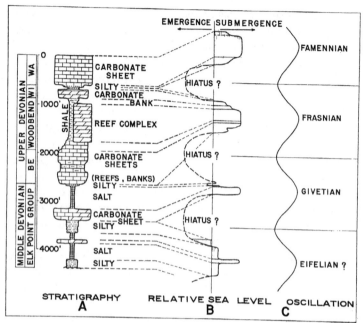

FIGURE 2. Interpretation of the changes in relative sea level during the deposition of Devonian beds in east-central Alberta.

during advanced regression and, ultimately, a hiatus in the stratigraphic sequence could develop as a result of non-deposition or erosion. The Upper Devonian sequence clearly incorporates two phases of sustained subsidence while a large part of the platform was inundated. The intervening phase of regression is marked by the presence of silt beds and one or several stratigraphic gaps.

An attempt can now be made to interpret the meaning of each unit in terms of sea level changes and of the relative duration of each depositional phase (Fig. 2). Obviously, the thickness of each unit is not proportional to the length of time required for its deposition. In order to rearrange the sequence of rock units into a series of time units, some graphic surgery is done to the vertical dimension of each unit to symbolize its duration. The result of this operation is, of course, highly subjective but not entirely devoid of reason (Fig. 2B). It can safely be assumed that during maximum transgression of the sea, the sedimentary record is more complete than during regression, and the importance of silty "breaks" and stratigraphic gaps has been stressed accordingly in the graph. In the chronological analysis, the missing parts of the stratigraphic succession are just as important as the preserved rocks we see. As a further aid in the adjustment of the stratigraphic column to a chronological sequence, it has been assumed that the epeirogenetic oscillations occurred with a regular rhythm.

This rhythm is represented by a single curve in Figure 2C. Some authors believe that the periods of predominant transgression lasted longer than the major regressions. This assumption is entirely speculative, although it must be correct in certain areas. In Figure 2C the rhythm of the oscillations is symbolized by a sine curve and it has been inferred that the duration of a stage in geological time can be scaled from the peak of one major regressive phase to the next. The stages may be of equal length, or they become progressively shorter in more recent geologic time.

Upon examination of the curve, certain well-known geological phenomena appear in logical succession. For example, the thick reef complexes may have a favourable start at the beginning of a transgressive phase and their growth is checked when the platform begins to emerge again. When regression sets in, the flat carbonate banks mark successive stages of retreat of the sea, but in places they are formed again when conditions are restored at the beginning of the following transgression.

It also stands to reason that those sediments which were deposited and preserved in a late phase of regression may be quite similar to those deposited during the early phase of the next transgression. Several examples could be given of the serious problems in mapping and correlation that arise as a result, and if the beds are not sufficiently well exposed, palaeontological data must be relied upon to resolve the difficulties.

The "explosive" change in fauna which accompanies some prominent stratigraphic breaks is more readily understood if the length of unrecorded geological time is taken into consideration.

Interpretive diagrams can be prepared for other Devonian successions in the demonstrated manner. Thus, representative sections in Saskatchewan, northern Alberta, the northern Mackenzie Mountains (N.W.T.), and the Arctic Islands were transposed into major oscillations of land and sea, and the diagrams are compared with the standard section of central Alberta in Figure 3.

A general consistency of the main changes in sea level is apparent, despite the uncertainty regarding correlation and age of some of these units. Different interpretations and improvements are definitely possible. The present analysis, however, does not primarily concern problems of correlation of lithologic units, but the possibility is investigated of differentiating major epeirogenetic episodes in historical geology. Several of the stratigraphic gaps that are indicated in Figure 3 have long been recognized, while others are less apparent. The sequences in some areas may be either too complete or so incomplete that no clear record of differential movements of land and sea can be discerned. It is important, therefore, that successions of varied lithology are first considered.

Wide-spread changes in relative sea level are conspicuous in the protected shelf region, where the response of rock facies to environmental changes is most pronounced. In the northern region, on the other hand,

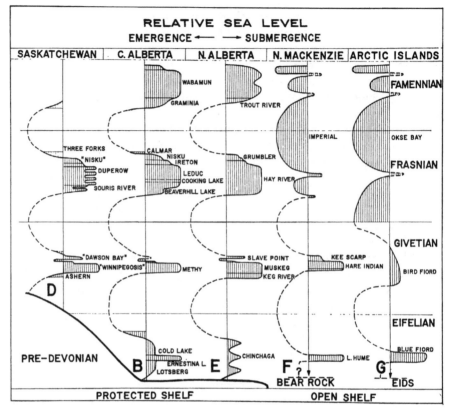

FIGURE 3.   Changes in relative sea level during the Devonian in several areas of western
Canada.

such stratigraphic changes are less obvious, but this may be due partly to a
lack of detailed observations. Further subdivision of the Imperial and Okse
Bay formations (of Upper Devonian age) is required, mainly because
the stratigraphic positions of marine tongues in these sandy formations have
not been accurately established.

In the Northwest Territories and on the Arctic Islands, beds of Middle
Devonian age are underlain by thick, poorly fossiliferous rock units. These
are the Bear Rock of the Mackenzie Basin and the Eids formation of the
Parry Islands Fold Belt. These formations are probably Lower Devonian
in age.

Our working hypothesis implies that the periodicity of epeirogenetic
movements of parts of the earth's crust may have a certain value in measuring geologic time. The validity of this basic assumption should be checked
against further information obtainable from stratigraphic and palaeontologic
studies and from absolute age determinations.

The suggested procedure presents difficulties from a practical point of
view. A considerable amount of regional geological information is needed

before it can be decided which stratigraphic breaks are most important with regard to the major oscillations.

Smaller oscillations or cycles, which are familiar to every geologist, occur in many rock series and consist of alternating laminations, beds, or groups of beds. These are of limited use in geochronological dating, because they are generally confined to one specific facies and cannot be traced regionally. Laminations may represent seasonal variations in sediment. The larger rhythmic units locally exceed a thickness of 100 feet. Many of these are probably the result of alternate destruction and construction of sedimentary barriers at the coastline during transgression, and they do not necessarily indicate oscillations of land or sea.

STRATIGRAPHY AND GEOCHRONOLOGY

In attempts to use stratigraphic data in geochronology it has generally been assumed that the thickest sedimentary succession (i.e. in the geosynclines) represents an almost complete record of geologic time. The consequences of this idea were examined by Schuchert (1931), who tried to resolve the average rate of sedimentation from the gross thickness of deposits formed during a known time interval, which had been determined by geochemical dating. The speculations of Schuchert constitute an interesting example of well-informed geological reasoning, but his premise appears to be untenable.

In a similar study, Kay (1955) stressed that the rate of sedimentation remains unknown. Some layers were deposited during a long time-interval, but other beds were formed almost at one instant. He stated also that: "A thickness of rock is not a steady accumulation of detritus, but the product of preservation of layers formed in short periods, separated by extremely long spaces of sparse or no deposition" (Kay 1955).

Most authors have abandoned all hope of using stratigraphic data to define time units. The thickness of a sediment is no parameter of geologic time unless more reliable data are obtained concerning rates of deposition. In other words, the sediments provide little direct information with regard to geochronology. If stratigraphic data are to be used it will be necessary, therefore, to consider indirect evidence such as the apparent rhythm in the succession of deposits.

A large number of geologists investigated the periodic occurrence of magmatic cycles, eustatic changes of sea level, ice ages, and particularly the orogenic disturbances. Most authors agree that a major rhythmic pattern of these phenomena exists and that changes have taken place in accelerated tempo.

Periodic deformations of the earth's crust probably occurred with chronological regularity if they originated mainly from physical and chemical processes in the interior of the earth. Bucher (1957) has shown that deformations within the inhomogeneous crust and the changes of sea level may be regarded as the complex response to alternating compression and tension due to internal changes. The important sedimentary rhythms stand out like the major and lesser pulses which appear among random "noise" on a seismic record.

Sloss (1963) has subdivided the stratigraphic column from latest Precambrian to Quaternary into six "stratigraphic sequences," which are rock-stratigraphic units of the highest rank. These sequences reflect main phases of interregional submergence and emergence in the history of the earth's crust. Optima of regression at the beginning of the Ordovician, Devonian, Pennsylvanian, Triassic, and possibly of the Cretaceous, seem to have occurred after periods of about 90 million years. Sloss based the stratigraphic sequences on physical relationships among rock units, however, and discussed their chronological significance only in general terms.

In most studies it has been attempted to detect a definite rhythm in the incidence of recurrent phenomena during geological time. In the present analysis of the Devonian system of western Canada this procedure was reversed. The regularity of major transgressive and regressive phases was tentatively accepted, and the rock units were appraised in terms of their position within the cycles. This approach offers the possibilities that certain stratigraphic observations may be used in the subdivision of time units which have been established by other means.

SELECTED BIBLIOGRAPHY

ANDRICHUK, J. M. (1951). Regional stratigraphic analysis of Devonian System for Wyoming, Montana, southern Saskatchewan and Alberta. Bull. Am. Assoc. Petrol. Geol., 35: 2368–408.

BAILLIE, A. D. (1953). Devonian names and correlations in the Williston Basin area. Bull. Am. Assoc. Geol., 37: 444–7.

BARRELL, J. (1917). Rhythms and the measurement of geologic time. Bull. Geol. Soc. Am., 28: 745–904.

BASSETT, H. G. (1961). Devonian stratigraphy, central Mackenzie River region, Northwest Territories, Canada. Geology of the Arctic, Proc. Intern. Symp. Arctic Geol., vol. 1, pp. 481–98.

BELYEA, H. R. (1963). Upper Devonian. Geological history of western Canada, chap. 6, pt. 2. Alta. Soc. Petrol. Geol. (in press).

BELYEA, H. R. and MCLAREN, D. J. (1962). Upper Devonian formations, southern part of Northwest Territories, northeastern British Columbia, and northwestern Alberta. Geol. Surv. Can., Paper 61-29.

BELYEA, H. R. and NORRIS, A. W. (1962). Middle Devonian and older Palaeozoic formations of southern District of Mackenzie and adjacent areas. Geol. Surv. Can., Paper 62-15.

BUCHER, W. H. (1957). Deformation of the Earth's Crust. London: Hafner.

BULLARD, E. C. (1944). Geological time. Mem. and Proc. Manchester Lit. and Phil. Soc., 86: 55–82.

CRICKMAY, C. H. (1957). Elucidation of some western Canada Devonian formations. Published by the author, Imperial Oil Ltd., Calgary.

——— (1960). The older Devonian faunas of the Northwest Territories. Published by the author, Imperial Oil Ltd., Calgary.

DOUGLAS, R. J. W. (1959). Great Slave and Trout River map-areas, Northwest Territories, 85 S 1/2 and 95 A, H. Geol. Surv. Can., Paper 58-11.

KAY, M. (1955). Sediments and subsidence through time. Geol. Soc. Am., Crust of the Earth, Spec. Paper 62, pp. 665–84.

McCONNELL, R. G. (1890). Report on an exploration of the Yukon and Mackenzie Basins, N.W.T. Geol. Surv. Can., Ann. Rept. 1888–1889, n.s., vol. 4, pt. D.

McLAREN, D. J. (1959). The role of fossils in defining rock units, with examples from the Devonian of western and Arctic Canada. Am. J. Sci., 257: 734–51.

——— (1962). Middle and early Upper Devonian Rhynchonellid brachiopods from western Canada. Geol. Surv. Can., Bull. 86.

McLAREN, D. J. and MOUNTJOY, E. W. (1962). Alexo equivalents in the Jasper region, Alberta. Geol. Surv. Can., Paper 62-23.

MOORE, R. C. (1952). Stratigraphical viewpoint in measurement of geologic time. Symposium on the measurement of geologic time. Trans. Am. Geophys. Union, 33: 149–203.

SCHUCHERT, C. (1931). Geochronology or the age of the earth on the basis of sediments and life. Bull. Natl. Res. Council, 80: 10–64.

SHERWIN, D. F. (1962). Lower Elk Point section in east-central Alberta. J. Alta. Soc. Petrol. Geol., 10(4): 185–92.

SLOSS, L. L. (1963). Sequences in the cratonic interior of North America. Geol. Soc. Am. Bull., 74 (2): 93–114.

UMBGROVE, J. H. F. (1939). On rhythms in the history of the earth. Geol. Mag., 76: 116–29.

——— (1942). The Pulse of the Earth. The Hague: Nyhoff.

WARREN, P. S. and STELCK, C. R. (1962). Western Canadian Givetian. J. Alta. Soc. Petrol. Geol., 10 (6): 273–91.

WHEELER, H. E. (1958). Time-stratigraphy. Bull. Am. Assoc. Petrol. Geol., 42: 1047–64.

GEOCHRONOLOGY OF PLUTONIC ROCKS IN TWO AREAS OF THE CANADIAN CORDILLERA

H. Gabrielse* and J. E. Reesor*

ABSTRACT

The evolution of plutonic rocks in north-central British Columbia, adjacent Yukon Territory, and southeastern British Columbia is discussed in the light of stratigraphy, K–Ar age determinations, and tectonic development. In both regions the plutonic complexes have developed over the span of an entire tectonic cycle, a minimum of 150 million years in duration. A combination of stratigraphic data and K–Ar ages suggests the development of plutonic rocks in a number of phases generally coincident with the principal tectonic episodes. K–Ar ages are in many instances successful in penetrating the effects of late phases, thereby recording earlier phases within the Mesozoic tectonic cycle and, in some cases, much earlier cycles of metamorphism and plutonic emplacement.

TWO EARLIER SUMMARIES of K–Ar ages from plutonic rocks in the Cordillera of western Canada have been published (Baadsgaard et. al. 1961; Muller, in Lowdon 1961). This paper attempts a further preliminary synthesis, using many new determinations, for two areas in British Columbia and the Yukon Territory. One was selected for availability of many K–Ar age determinations and considerable detailed information on the plutonic rocks, and the other for the availability of stratigraphic control (Fig. 1).

Figure 2 shows a plot of age determinations in southern British Columbia compared with a plot of all other determinations from British Columbia and Yukon Territory. All ages plotted here from the two areas under discussion are listed in Table I (Appendix). All others have been obtained from Lowdon (1960, 1961, 1963; and Lowdon et al. 1963) and Baadsgaard et al. (1961). It is evident from these plots that Mesozoic ages are concentrated in a series of maxima roughly equivalent in number and position for both southern British Columbia and for the Canadian Cordillera as a whole. From the age patterns, and from broad compositional, textural, and structural similarities, it is possible to suggest a common pattern of Mesozoic plutonism throughout the Canadian Cordillera.

The significance of the maxima of Fig. 2 can only be evaluated sensibly if the following factors are kept in mind.

1. The complex succession of structural events within the Mesozoic cycle of deformation.

2. The associated complex evolution of the plutonic rocks throughout the development of the mobile belt during the Mesozoic.

*Geologist, Geological Survey of Canada.

3. Complex, gradual cooling of emplaced masses during uplift and unroofing.

Any one or, more probably, all of these factors affect the resulting isotopic age measurements. Thus, unravelling the precise timing of plutonic and associated tectonic activity in a mountain belt is not a simple matter of

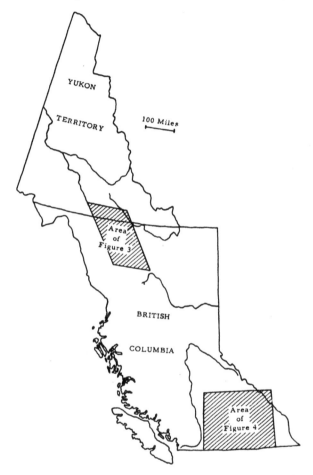

FIGURE 1. Key map showing location of areas under discussion.

obtaining scattered age determinations from granitic or metamorphic rocks. On the contrary the significance of individual ages can be judged only by integrating information of many kinds from many localities.

The first step in relating plutonic and tectonic events is to examine the stratigraphic record and the associated granitic rocks in a region in which the stratigraphic relationships are reasonably well established. In British Columbia the Mesozoic record is perhaps best exposed in the north (see Figs. 1 and 3).

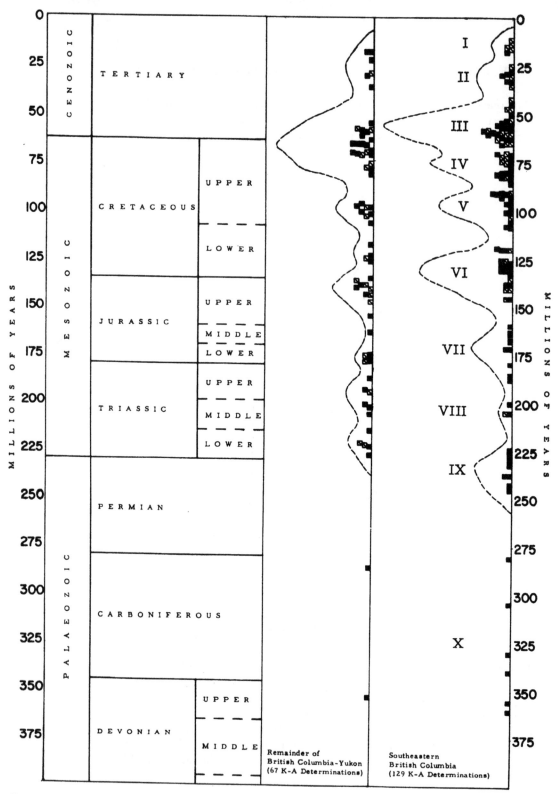

FIGURE 2. Plot of 196 K–Ar age determinations from British Columbia and Yukon Territory. Each square represents one determination: open squares, metamorphic rocks; closed squares, granitic rocks. Time-scale modified after Kulp (1961). Roman numerals refer to age maxima plotted on Figure 4.

PLUTONIC ROCKS IN PARTS OF NORTHERN BRITISH COLUMBIA
AND SOUTHERN YUKON

The influence of two prominent belts of granitic and metamorphic rocks on the development of the major structural and stratigraphic elements in northwestern British Columbia and southwestern Yukon Territory has been discussed (Gabrielse and Wheeler 1961). On evidence obtained mainly from adjacent, largely unmetamorphosed stratified rocks it was shown that the crystalline belts have had a long and active history. This conclusion has been strengthened by additional information from isotope ages of plutonic rocks and by further field work. The following discussion deals with the eastern crystalline belt (Cassiar crystalline belt) and adjacent rocks, exposed mainly in the Cassiar Mountains (Fig. 1).

The largest body of granitic rocks in the Cassiar crystalline belt is the Cassiar batholith. This batholith includes a variety of granitic rock types ranging from gabbro (minor) through quartz diorite, granodiorite, quartz monzonite, and granite. Although general areas underlain by the various types have been recognized, the relations between them have not been worked out. Further work may well show that the Cassiar batholith is composite in that it includes granitic bodies of several compositions and ages. Such a situation has been suggested for granitic rocks in the Tuya–Teslin area southwest of the Cassiar batholith and southeast of Teslin Lake (Watson and Mathews 1944). There, granite and granodiorite of the Glundebery batholith are intrusive into diorite of the Christmas Creek batholith. Within the Glundebery batholith is a small body of graphic and miarolitic granite believed to be intrusive into the larger mass.

Roots (1954) noted that melanocratic rocks, containing little or no quartz, were the oldest phases of the Hogem batholith exposed in the Omineca Mountains southeast of the region discussed in this paper. The remainder of the Hogem batholith is described as a composite body composed of a number of individual igneous bodies generally separated by intrusive contacts. A general sequence of intrusion, with but few minor reversals, is from diorite through quartz diorite and granodiorite to adamellite-granite.

Other bodies of granitic rocks near the Cassiar batholith, including the Marker Lake, Seagull, and Hotailuh batholiths, appear to have fairly uniform composition.

Except near contacts with granitic rocks the metamorphic rocks of the eastern crystalline belt are of uniformly low grade and consist mainly of mica-quartz schists and gneisses containing more or less chlorite and albite. Granitic gneisses have been developed locally near the north end of the Horseranch Range.

Although the regionally metamorphosed rocks are in a general way spatially related to granite rocks they are clearly transected by them. The metamorphic rocks are for the most part, therefore, not simple products of contact metamorphism related to the emplacement of granitic plutons but are products of a relatively early episode of plutonic activity that pre-dates the intrusion of the bulk of the granitic rocks. All of the granitic bodies within and near the Cassiar crystalline belt are clearly intrusive. Only in the granitic gneiss terrain that underlies a small area near the north end of Horseranch Range does granitization appear to have played a significant role.

GENERAL GEOLOGY

The Cassiar crystalline belt is flanked to the east and northeast by rocks of mainly Palaeozoic age and to the southwest by rocks of Palaeozoic and Mesozoic age (see Fig. 3). Miogeosynclinal rocks to the east range in age from Proterozoic to Late Devonian. East of Dease River regionally meta-morphosed Cambrian and possibly older rocks occupy a faulted, doubly plunging anticlinorium within the miogeosynclinal terrain.

The non-volcanic strata described above are overlain by a very thick sequence of rocks including abundant volcanics and ultramafics. This eugeosynclinal assemblage is mainly of Devono-Mississippian age but locally includes Permian and Upper Triassic rocks. Locally, Middle Mississippian limestones unconformably overlie the Devono-Mississippian sequence.

Southwest of the Cassiar crystalline belt a thick eugeosynclinal sequence, possibly ranging in age from Devono-Mississippian to Permian, is overlain by a thick assemblage of Mesozoic and Tertiary strata that include volcanic and non-volcanic units. A summary of the stratigraphy is given in Table II. The rocks are grouped in depositional sequences that exhibit considerable variations within themselves but can be readily separated from older and younger sequences.

METAMORPHIC ROCKS

Several ages are represented by the regionally metamorphosed rocks in the Cassiar crystalline belt. Metamorphic rocks near Teslin Lake have been mapped by Mulligan (1955) as Mississippian or earlier and by Poole et al. (1960) as Upper Devonian and Lower Mississippian. Rocks in the meta-morphic terrain surrounding the Marker Lake batholith are mapped by Poole et al. (1960) as Lower Cambrian and (?) earlier. Watson and Mathews (1944) were unable to date the metamorphic rocks northwest of Dease Lake but pointed out that part of the formation (Oblique Creek Formation) represented a metamorphosed group of rocks similar to that of the Kedahda Formation of Permian age. Further work has not solved the problem but has emphasized the importance of the relationship between the Kedahda Formation, Oblique Creek Formation, and Sylvester Group in the stratigraphic dating of regional metamorphism in the Cassiar crystal-line belt. The Oblique Creek Formation can be traced southeasterly into Dease Lake and Cry Lake map-areas (Gabrielse and Souther 1962;

Gabrielse 1962*a*) to northeast of Dease Lake where the metamorphic rocks lie well within the Cassiar batholith. Metamorphic rocks farther east and southeast represent non-volcanic Lower Palaeozoic strata.

A thick sequence of regionally metamorphosed strata outcrops northeast and east of the Cassiar batholith southeast of Dall Lake. This sequence is probably mainly or entirely correlative with the Ingenika Group (Roots 1954), in part of Lower Cambrian age. These rocks include much chlorite schist and in gross aspect differ considerably from Lower Cambrian strata exposed to the northeast in the Cassiar and Rocky Mountains (Gabrielse 1962*b*). The general trend of facies belts in this part of northern British Columbia suggests that the "Ingenika facies" northwest of Dall Lake was restricted to the Cassiar crystalline belt or even farther to the southwest.

GEOCHRONOLOGY OF PLUTONIC ROCKS
IN AND NEAR THE CASSIAR CRYSTALLINE BELT

Suggested times of plutonic activity in the Cassiar crystalline belt are shown in Table II. These times are indicated by a combination of stratigraphic data and K–Ar isotopic ages on plutonic rocks. Because relatively few isotopic ages are yet available for the region, much of the emphasis in the following discussion is placed on stratigraphic evidence and these data are presented first.

Stratigraphic Evidence on Age of Plutonic Activity

As mentioned above, the formation of a zone or zones of regionally metamorphosed rocks occurred relatively early in the plutonic history. In general an initial widespread metamorphism occurred either in late Palaeozoic or early Mesozoic time. At present, information is not available to date this metamorphism more closely. The belt of Devono-Mississippian rocks east of the Cassiar batholith (Sylvester Group) is transected by the granitic rocks and forms part of a regionally metamorphosed terrain near Teslin Lake (Gabrielse, in press; Poole *et al.* 1960; Mulligan 1955). It is thus clear that at least part of the regional metamorphism is post-Devono-Mississippian. If the Kedahda Formation is in part or entirely of Permian age and if the metamorphic Oblique Creek Formation represents correlative strata, then a post-Permian, probably lower Mesozoic metamorphism is indicated.

In Wolf Lake map-area, southern Yukon Territory, a conglomerate contains fragments of metamorphic rocks and Permian fusulinids (Poole, pers. comm., 1960). If the fusulinids date the conglomerate, at least some metamorphism must have taken place in late Palaeozoic time. Perhaps it is also significant that only one occurrence of Permian rocks is known east of the Cassiar crystalline belt and in a general way Permian rocks thin easterly towards the belt (Gabrielse and Souther 1962). In addition, Pennsylvanian rocks have not been recognized in the region under discussion.

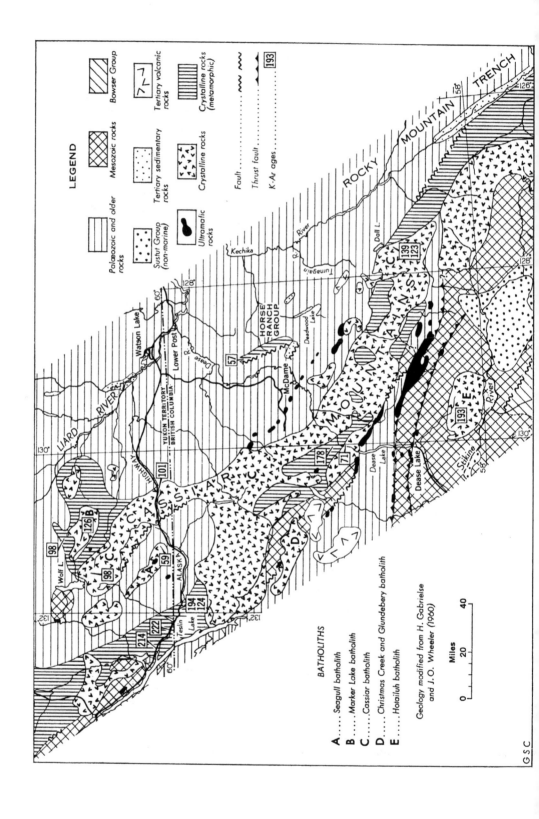

LEGEND

Palaeozoic and older rocks

Sustut Group (non-marine)

Bowser Group

Mesozoic rocks

Tertiary sedimentary rocks

Tertiary volcanic rocks

Ultramafic rocks

Crystalline rocks

Crystalline rocks (metamorphic)

Fault ....

Thrust fault ....

K-Ar ages 193

BATHOLITHS

A ..... Seagull batholith
B ..... Marker Lake batholith
C ..... Cassiar batholith
D ..... Christmas Creek and Glundebery batholith
E ..... Hotailuh batholith

Geology modified from H. Gabrielse and J.O. Wheeler (1960)

Miles

0   20   40

GSC

ROCKY MOUNTAIN TRENCH

CASSIAR MOUNTAINS

LIARD RIVER

Watson Lake

Lower Post

YUKON TERRITORY
BRITISH COLUMBIA

Wolf L.

Teslin Lake

ALASKA
HIGHWAY

Dease Lake

Stikine R.

HORSE RANCH GROUP

Kechika R.

Turnagain R.

Dall L.

Deadwood Lake

McDame

Dease River

TABLE II

Generalized Table of Formations and Suggested Episodes of Plutonic Activity in and near the Cassiar Crystalline Belt

Roman numerals represent time groups taken from Figure 2

PERIOD		GROUP OR FORMATION	DOMINANT LITHOLOGY		PLUTONIC ACTIVITY
CENOZOIC				III, IV ←	KASTBERG INTRUSIONS SEAGULL BATHOLITH HORSERANCH GROUP (phase of regional metamorphism (?) CASSIAR BATHOLITH (phase of (?)
		SUSTUT GROUP	Non-marine sandstone, shale, conglomerate 5,000'		
CRETACEOUS	U				CASSIAR BATHOLITH (phase of)
			Unconformity	V ←	
	L	BOWSER GROUP	Mainly marine greywacke, shale, pebble conglomerate 10,000'	VI ←	CASSIAR BATHOLITH (phase of) MARKER LAKE BATHOLITH
	U				
JURASSIC	M		Unconformity		
	L	UNNAMED GROUP	Greywacke, shale, conglomerate 7,000'		
			Unconformity	VIII ←	HOTAILUH BATHOLITH CASSIAR BATHOLITH (phase of (?)
	U	UNNAMED GROUP	Volcanic rocks, clastic rocks, limestone 5,000'		
			Unconformity		
TRIASSIC	M	UNNAMED GROUP	Chert, argillite, greenstone, conglomerate, age uncertain 1,000'		
	L		Unconformity	IX ←	CASSIAR CRYSTALLINE BELT (regional metamorphism)
		TESLIN GROUP	Limestone 500'		
PERMIAN			Conformable (?)		
		KEDAHDA FORMATION	Chert, argillite, slate, quartzite, greenstone, limestone 7,500'		
PENNSYLVANIAN		Relations unknown Kedahda Formation may include rocks older than Permian		X ←	CASSIAR CRYSTALLINE BELT (possible metamorphism)
MISSISSIPPIAN	U M	NIZI FORMATION	Limestone 1,000'		
	L		Unconformity		
DEVONIAN	U	SYLVESTER GROUP	Chert, argillite, slate, chert arenite, conglomerate volcanic rocks, ultramafic rocks 15,000'		
		Disconformity (?) near Cassiar batholith marked unconformity near Rocky Mountain Trench			
		UPPER AND PRE-UPPER DEVONIAN Limestone, dolomite, sandstone, shale			

A complicating factor in determining the maximum age of regional metamorphism in the Cassiar crystalline belt is that the southwestern and western limits of the metamorphic terrain are unknown. Devono-Mississippian rocks of the Sylvester Group are relatively unmetamorphosed immediately northeast of the Cassiar batholith along Dease River and near Seagull batholith. The northeastern limit of the metamorphic terrain is therefore fairly well established. To the southwest, however, it is not known whether the relatively unmetamorphosed Permian rocks lie outside of the crystalline belt or overlie it. In the latter case it could be argued that the Oblique Creek Formation is of Devono-Mississippian age and that the assemblage might be correlated with the Sylvester Group rather than with the Kedahda Formation as these units are very similar in general aspect.

Mulligan (in Lowdon 1963, pp. 26, 27) noted the contrast in degree of metamorphism between Mississippian strata and overlying Permian and Mesozoic rocks in the Teslin area. Such a situation, of course, would support the contention that Permian rocks are, at least in part, younger than the regional metamorphism.

Upper Triassic rocks are nowhere significantly metamorphosed, even adjacent to regionally metamorphosed rocks. On this basis regional metamorphism within the Cassiar crystalline belt is regarded as pre-Upper Triassic.

Little evidence is available as to the age of metamorphism in the terrain southeast of Dall Lake except that it was post-Lower Cambrian. Roots (1954) suggested a post-Lower Cambrian, probably pre-Mississippian, granitization episode that produced the Wolverine Complex southeast of the region described in this paper. He based this conclusion on the presence of granodiorite and micaceous, gneissic quartzite boulders resembling rocks of the Wolverine Complex, in a conglomerate contained in an assemblage of Mississippian to probably Permian age.

In summary, stratigraphic evidence within the region described indicates that an episode or episodes of regional metamorphism affected rocks of the Cassiar crystalline belt during post-Devono-Mississippian and pre-Upper Triassic time. It should be noted that during this interval there are two conspicuous breaks in the stratigraphic record evidenced by the absence of recognized Pennsylvanian and Lower Triassic strata.

No strata of early Triassic age have been found in the northwestern Cordillera. West of Dease Lake rocks of mid-(?)Triassic age include basal conglomerate containing fragments of Permian limestone. Deposition of wide-spread Upper Triassic (Karnian and Norian) rocks was followed by another disturbance that resulted in the absence of uppermost Triassic (Rhaetian) rocks in the region and evidently throughout the entire northwestern Cordillera.

The first significant influx of granitic material into the sedimentary record came in early Jurassic time. Southwest and southeast of Dease Lake Lower Jurassic conglomerates that unconformably overlie Upper Triassic rocks contain pebbles and cobbles of granitic rocks. The granitic rocks are distinctive and can be correlated with similar rocks that occur in plutons intrusive into Upper Triassic (Karnian and Norian) rocks. The largest of these plutons is the Hotailůh batholith, a relatively homogeneous, even-grained, pink-weathering, hornblende granodiorite. Similar rocks, though highly sheared, occur to the southwest in the Grand Canyon of Stikine River. The Hotailuth batholith thus appears to have been emplaced during latest Upper Triassic (Rhaetian) time.

A regional unconformity marks the base of a sequence of late Jurassic and early Cretaceous sediments deposited in the Bowser Basin, but whether the related uplift is associated with plutonic activity is not known. No fossils of Bathonian age (mid-Middle Jurassic) have been found in the north-western Cordillera of Canada (H. W. Frebold, pers. comm., 1963).

In Omineca Mountains a conglomerate (Uslika Formation) of probable Lower Cretaceous age (Roots 1954) contains pebbles, cobbles, and boulders of plutonic rocks. Significantly, perhaps, diorites, syenodiorites, and quartz-poor porphyritic granodiorite are proportionally more abundant in the conglomerate than are leucocratic quartz-rich granitic rocks, although the latter are now considerably more abundant in the Hogem batholith at the present level of erosion.

Another regional unconformity occurs at the base of late Cretaceous and early Tertiary non-marine sedimentary rocks (Sustut Group). Typical biotite–quartz–monzonite and biotite–granodiorite of the Cassiar batholith cut Lower Jurassic strata, and boulders of these lithologies are locally abundant in conglomerates of the Sustut Group (Lord 1948; E. F. Roots, pers. comm., 1958). A wide-spread emplacement of quartz monzonite and granodiorite in the Cassiar crystalline belt probably took place, therefore, during the post-early Jurassic and pre-Upper Cretaceous interval. South-west of the Cassiar batholith near Dease Lake a broad belt of Lower Jurassic and older rocks is characterized by structures reflecting forces directed to the southwest. These structures, including two major thrust faults, are believed to be related to the emplacement of a great volume of granitic rocks into the Cassiar crystalline belt. Significantly, this distinctive pattern of deformation does not appear in overlying strata of the Sustut Group to the southeast.

In McConnell Creek map-area southeast of the region described herein strata of the Sustut Group have been intruded by porphyritic granitic rocks of the Kastberg Intrusions (Lord 1948). These rocks range in composition from granodiorite to quartz diorite and are characterized by being pre-dominantly finer-grained than most granitic rocks in the area, and by being, in most cases, thoroughly weathered near the surface. Lord suggests that the porphyries were most probably emplaced in early Tertiary time.

In southern Yukon Territory, Poole (1956) described a body of miaro-litic, tourmaline-bearing, leucoquartz monzonite (Seagull batholith) that cuts structures believed related to the emplacement of the Cassiar batholith.

As noted previously, miarolitic granite southwest of Cassiar batholith appears to represent the latest phase of plutonic activity in that area (Watson and Mathews 1944).

In summary, stratigraphic evidence indicates that the area including the Cassiar crystalline belt underwent a phase of regional metamorphism in late Palaeozoic or early Mesozoic time and at least three episodes of granitic intrusion during late Triassic (Rhaetian), post-early Jurassic and pre-late Cretaceous, and post-earliest Tertiary times. The significance of available K–Ar isotopic ages of plutonic rocks in the region can now be examined to see how they fit the information obtained from stratigraphy and how they might contribute to a fuller understanding of the plutonic history.

K–Ar Isotopic Ages of Plutonic Rocks

Isotopic ages of plutonic rocks from and near the Cassiar crystalline belt are included in Table I and Figure 3. Further information on these determinations is contained in publications by Lowdon (1961, 1962, and 1963; and Lowdon *et al.* 1963) and in a paper by Baadsgaard *et al.* (1961).

It is obvious that isotopic ages of granitic and metamorphic rocks in this area are still much too few to permit a comprehensive evaluation of their significance. In a general way, however, the available determinations support the conclusion obtained from stratigraphic data that plutonic activity embraced essentially all of the Mesozoic and extended into the Tertiary.

The oldest dates (222 and 214 m.y.) were obtained from regionally metamorphosed rocks near Teslin Lake. The suggested early Triassic age is in accord with the absence of early Triassic strata, the absence of regionally metamorphosed Mesozoic rocks, and the transecting of the metamorphic rocks by later granitic bodies. Two further ages (194 and 124 m.y.) obtained on a muscovite–biotite pair from similar rocks farther southeast in the metamorphic belt are perhaps significant in the light of the two older ages. The age of 194 m.y. obtained on the muscovite is possibly slightly low. This is supported by the date provided by the biotite (124 m.y.), indicating a considerable loss of argon from this mineral relative to the muscovite. An interpretation of this pair of ages is that, locally, the regionally metamorphosed rocks underwent a mild metamorphism or reheating that brought about a substantial argon loss from the biotite and possibly a minor argon loss from the muscovite. This reheating could be related to a later emplacement of granitic rocks.

The date of 178 m.y. on muscovite from schists of the Oblique Creek Formation substantiates the field relationships in that the metamorphism of these rocks generally pre-dated emplacement of most of the granitic rocks. This date must also be considered a minimum age in view of the proximity of later granitic intrusions.

The oldest isotopic age obtained on granitic rocks so far is 193 m.y. on biotite from the Hotailuh batholith. This date fits very well with field evidence indicating an emplacement of granitic rocks during late Triassic (Rhaetian) time. More isotopic ages are required from the Hotailuh batholith, however, to determine whether the apparent homogeneity of this body results in concordant ages.

Isotopic age dates from granitic rocks of the Cassiar and Marker Lake batholiths range from 139 m.y. to 71 m.y. These ages are difficult to assess because precise stratigraphic information is lacking. The range expressed by the dates is believed real, however, in the sense that several episodes of intrusion are probably represented. A large mass of hornblende-rich diorite and quartz diorite occurs within the southeasternmost area of granitic rocks shown in Figure 3. The rocks are faulted and traversed by numerous shears and epidotized fractures. In this respect they contrast with the much fresher appearing quartz monzonites of the Cassiar batholith. These dioritic rocks may well be older than any granitic rocks of the Cassiar batholith so far sampled for isotopic age determination. The two oldest ages for the Cassiar batholith (139 and 123 m.y.) and an age for the Marker Lake batholith (126 m.y.) fall near the Jurassaic–Cretaceous boundary (see Fig. 2). It is indicated in Table II that at this time sedimentation may have been continuous in the Bowser Basin. Perhaps one can accept the concept of plutonic activity in the Cassair crystalline belt concomitant with sedimentation in an adjacent area. On the other hand, it has been shown that in all other cases the times of plutonic activity, well documented by stratigraphy, are also times of regional uplift. The problem may be resolved by detailed work on the stratigraphy of the Bowser Basin. Early Cretaceous uplift and erosion is evidenced by the Uslika Formation in the Omineca Mountains (Roots 1954).

Ages of 101 and 98 m.y. from the Cassiar batholith in Yukon Territory would fit a mid-Cretaceous age consistent with stratigraphic evidence. Metamorphic rocks near the Marker Lake batholith must also have been recrystallized at this time (98 m.y.).

The indicated age for the Seagull batholith (59 m.y.) agrees with the suggestion by Poole (1956) that this intrusion is younger than the nearby Cassiar batholith. An age of 71 m.y. from coarsely crystalline quartz monzonite included in the Cassiar batholith near the north end of Dease Lake requires confirmation as stratigraphic evidence on the time that these rocks were emplaced is lacking. In Table II the Seagull batholith is suggested as being of post-Sustut age but this is based only upon the isotopic age combined with meagre stratigraphic evidence outlined above.

An early Cenozoic age (57 m.y.) on muscovite from the Horseranch Group fits in a general way the relatively young ages so far obtained on rocks of the Wolverine Complex to the southeast—see Muller (*in* Lowdon 1961) and Lowdon *et al.* (1963). The significance of these ages is not known. A similar situation exists with isotopic ages of parts of the Shuswap Complex in southern British Columbia, many of which seem inexplicably young.

SUMMARY OF PLUTONIC ACTIVITY

Table II presents a tentative summation of the data discussed above. As a word of caution one must constantly keep in mind that although most of the isotopic ages can be fitted into the stratigraphic column fairly well the active history of the mobile area offers considerable leeway for this purpose. For instance, although the oldest isotopic ages for metamorphic rocks in the Cassiar crystalline belt indicate an early Triassic age, the ages must still be considered as being minimum values. It would be premature, therefore, to rule out the possibility of a pre-Permian metamorphism simply because two of the available isotopic ages accommodate the stratigraphic evidence for an early Triassic age. On the other hand, it might be unwise to dismiss the younger isotopic ages of the regionally metamorphosed rocks from the Cassiar crystalline belt as simply reflecting the loss of argon from the micas. The younger ages may be significant in that they may accurately depict times of recrystallization, reheating, or structural dislocation of the previously metamorphosed rocks. This is suggested by the close correlation of the younger ages of metamorphic rocks with the various ages of granitic rocks.

With the many limitations in mind a few generalities may be made on the plutonic history of the region near, and including, the Cassiar crystalline belt.

1. The plutonic "cycle" has had a long history extending possibly from late Palaeozoic through Mesozoic and into the Tertiary.

2. In a general way episodes of significant plutonic activity have coincided with marked regional breaks in the stratigraphic column. The problem of possible granitic emplacement in the Cassiar crystalline belt coinciding with deposition in the Bower Basin has been discussed above.

3. Throughout the "plutonic cycle" there was a general succession from hornblende-bearing diorites and granodiorites through biotite-bearing quartz monzonites to granites. During the Cenozoic, porphyritic and finer-grained quartz diorites and granodiorites were emplaced locally.

4. Episodes of plutonic activity occurred in pulses at intervals of about 30 m.y. and appear to have been of regional significance.

SOUTHERN BRITISH COLUMBIA

The foregoing discussion has indicated the principal patterns of age results in a region of reasonably well established stratigraphic relationships. Mesozoic stratigraphic patterns in this part of southern British Columbia are not easily deciphered, partly because Mesozoic strata occur in narrow, highly deformed, belts, but partly because vast areas expose older rocks ranging in age from late Precambrian to Palaeozoic. Much of the region is underlain by granitic or high-grade metamorphic rocks in which no precise stratigraphic or faunal succession will ever be established. It is of great importance, therefore, to establish the usefulness of age determinations in such a terrain. The problem becomes essentially one of understanding the succession and patterns of plutonism, for this may be expected to reflect the over-all tectonic development of the mobile belt.

Figure 4 shows the main granitic masses as well as the Shuswap Metamorphic Complex in southern British Columbia. In order to understand better the broad relationships of the granitic masses to one another and to the mountain belt as a whole it is necessary to review briefly the principal geological patterns in this region.

GEOLOGICAL SUMMARY

Belts of distinctive sedimentary successions showing a characteristic structural style and grade of regional metamorphism are generally parallel with the northward trend of both the line of individual plutons and the Shuswap Metamorphic Complex. Along the eastern fringe of this mountain belt lie the Rocky Mountains, composed principally of Palaeozoic rocks exhibiting little regional metamorphism, but dominated by folded and thrust structures of relatively competent rock units on a grand scale. Few granitic rocks are found in this belt.

The Rocky Mountain Trench divides the Rocky Mountains and the Purcell – northern Selkirk arc. West of the Trench, the rock successions consist principally of clastic sediments of late Precambrian age in the southern Purcell Mountains followed by successively younger, latest Precambrian – lower Palaeozoic sediments of the northern Purcell and Selkirk Mountains. Rocks of this belt are complexly folded along generally northward trends that conform to the general trend of Purcell–Selkirk arc. Metamorphism gradually increases westward, though localities within the belt may reach a sillimanite or kyanite grade of regional metamorphism, e.g. central Kootenay Lake (Crosby 1960) or the northern Selkirks, south of Kinbasket Lake (Wheeler 1961).

This region is characterized by a succession of small, discrete, intrusive, plutonic masses, arranged in a broad arc, symmetrical with the general arcuate pattern of the fold trends. However, each individual mass, with few exceptions, lies athwart and often greatly modifies the northward-trending folds.

Within the concave side of this arc, in the southern Selkirk Mountains, lies a complex belt of Mesozoic volcanics and fine clastics. It is fringed on the east by a narrow belt of upper Palaeozoic rocks. This belt is dominated by two large, subconcordant, granitic plutons, the Nelson and Kuskanax batholiths. It is characterized as well by complex folding along the Kootenay arc that involves not only the younger Palaeozoic and early Mesozoic rocks, but also the older Palaeozoic and latest Precambrian rocks, at least

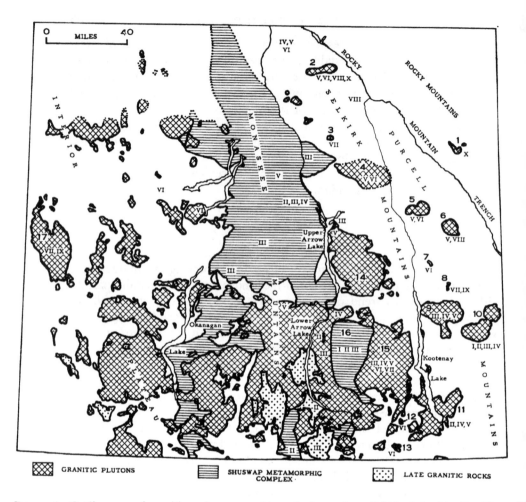

FIGURE 4. Outline map of granitic and metamorphic rocks in southern British Columbia. See **Figure 2** for age groups shown here by Roman numerals.

Location key for individual masses	Time groups from Figure 2
1. Ice River Complex	I. 10–18 m.y.
2. Adamant batholith	II. 25–35 m.y.
3. Fang stock	III. 40–65 m.y.
4. Battle batholith	IV. 70–85 m.y.
5. Bugaboo batholith	V. 90–110 m.y.
6. Horsethief batholith	VI. 117–145 m.y.
7. Glacier stock	VII. 160–185 m.y.
8. Toby stock	VIII. 195–205 m.y.
9. Fry Creek batholith	IX. 225–245 m.y.
10. White Creek batholith	X. 281 m.y.
11. Bayonne batholith	304–360 m.y.
12. Porcupine Creek stock	
13. Lost Creek stock	
14. Kuskanax batholith	
15. Nelson batholith	
16. Valhalla Complex	
17. Guichon batholith	

of the western part of the Purcell Mountains. These rocks are considered to be folded in two phases along axes that trend generally along the Kootenay arc (Fyles 1962). Rocks in this belt of upper Palaeozoic or younger age appear to be little metamorphosed except in the vicinity of granitic plutons. Rocks of early Palaeozoic age are somewhat more metamorphosed, and locally may reach a sillimanite grade of metamorphism.

The Shuswap Metamorphic Complex forms a broad, generally northward trending belt along the central axis of this part of the western Cordillera in southern British Columbia. It is about 100 miles wide at the southern end of its exposure, narrows abruptly at the 50th parallel of latitude to 60 miles wide, then plunges gradually northward beneath rocks of lower metamorphic grade just north of latitude 52° (the northern limit of Fig. 4). The Shuswap Metamorphic Complex is characterized by a high grade of regional metamorphism almost invariably in the sillimanite-almandine subfacies of the amphibolite facies. Characteristic structural style in the Shuswap consists of an early east–west lineation marked by both mineral orientation and axes of folds. The early east–west isoclinal, recumbent folds are apparently related to major east–west warps, at least along the eastern limit of the Shuswap, that occur at approximately 40 to 50 mile intervals. These early folds and the associated lineation are oriented normal to the general trend of the Shuswap, as well as normal to the trend of folding in the associated belts east or west of the Shuswap Metamorphic Complex. This early plastic deformation has taken place at depth as shown by the character of the folding, associated migmatization, and high grade of regional metamorphism (Jones 1959; Reesor 1963).

At least one major, later tectonic event has affected the earlier structural and metamorphic patterns of the Shuswap. In most localities this has resulted in a penetrative redeformation of the early east–west warps, folds, and lineations, resulting in a superimposed set of northward-trending small-scale structures, both folds and lineations, as well as major northward-trending warps. Thus, the general northerly Cordilleran trend was superimposed for the first time during this episode. The resulting major structural pattern is a succession of domal structures formed at a somewhat shallower zone than the early east-west structures, such that easiest relief of movement has been vertically upward, and domal cores have become diapirs. (Reesor 1963). These structures have also been formed at depth, and recrystallization in most localities has been intense. Even in localities at which the latest pattern of deformation is not megascopically evident, quartz microfabric diagrams show an incipient pattern similar to that in other zones clearly affected by the late deformation (Reesor 1963).

The Shuswap Metamorphic Complex is not considered here to be a group of rocks of a particular age; it may well contain rocks ranging in age from Precambrian to early Mesozoic. It is marked solely on the basis of its structural pattern and high amphibolite grade of regional metamorphism. The Complex is unconformably overlain by rocks of Oligocene(?) age (Jones 1959, p. 52).

With the exception of the region north of Revelstoke, the Shuswap is bounded on the east by belts of rocks that are youngest (Mesozoic) nearest the metamorphic complex and oldest (late Precambrian) farther east near the Rocky Mountain Trench. In contrast, successive belts west of the complex are apparently oldest (Late Precambrian?, early Palaeozoic?) near the complex and become progressively younger farther away (Cretaceous). This broad pattern could be interpreted to mean that rocks of all ages from Precambrian to at least early Mesozoic are involved in at least part of the structural deformation and metamorphism considered to be characteristic of the Shuswap. However, much further work is necessary to define clearly the relationships of the Shuswap Metamorphic Complex to this Mesozoic mountain belt.

In considering the granitic rocks from all structural and metamorphic zones in southern British Columbia on the basis of local relationship to structure and to cross-cutting relations within individual plutons a granitic succession may be identified without reference to age determinations. Briefly, early granitic types consist of hypersthene monzonite and quartz diorite, followed by mafic-rich granodiorite that is, in turn, followed by leucocratic granodiorite, quartz monzonite, and leucogranite. The latest granitic succession in the region consists of syenodiorite, pyroxene monzonite, minor syenite, and much hornblende quartz monzonite and granite. The relation of the soda-rich pyroxene and hornblende leucomonzonite and leucoquartz monzonite of Kuskanax batholith to this succession is not known.

Within this geologic framework, as briefly outlined here, we may now consider the associated patterns of plutonism and age determinations in southern British Columbia.

AGE DETERMINATIONS

Precambrian and Palaeozoic Ages

Precambrian ages have been obtained from basic, granitic, and metamorphic rocks from the Purcell System in the southern Purcell Mountains. The principal group of ages ranges from 600 m.y. to over 800 m.y. and has been obtained from hornblende and biotite from basic sills (Hunt 1962), and on muscovite from granite and metamorphic rocks (Leech 1962). Hunt (op. cit.) has reported ages as old as 1580 m.y. on hornblende from basic sills.

The principal age group, 600–800 m.y., is considered to represent a minimum age for a pre-Windermere orogenic episode long evident from stratigraphic data in this region (Leech, op. cit.). However, it is surprising, in view of the long and complex subsequent history of this region, that any ages earlier than the Mesozoic have been preserved. That they have been preserved has perhaps been the result of the local environment and the nature of the subsequent deformation. The pluton of Hell-Roaring Creek (Leech 1962) and the metamorphic rocks of Mathew Creek lie within a great succession of competent lower Purcell strata that presumably preserved localities of earlier metamorphism or earlier granite plutons and Moyie sills as essentially passive elements, thus preventing a penetrative redeformation or metamorphism. Nevertheless, a great spread of hybrid ages is available from different localities on different minerals. Some show clearly the effect of hybridization (Hunt 1962; Leech, *in* Lowdon 1963). Can it then be assumed that those determinations that preserve the oldest ages preserve in fact the "absolute" age of a specific orogenic or metamorphic event? Corroboration of specific events must be sought from determinations by other methods of isotopic age determinations, preferably zircon or whole-rock Rb–Sr isochrons. Nevertheless, though the K–Ar ages may be somewhat less than precise, they confirm the existence of events much earlier than the culminating Mesozoic orogeny in this region, and in fact add much to our knowledge of the intensity of the pre-Windermere orogenic event and give at least a minimum "absolute" age for that event.

The mid-Palaeozoic ages obtained from Ice River Complex (see Table I and Fig. 4) in the Rocky Mountains again demonstrates the survival of older ages in spite of intense later structural deformation (Laramide). Ages in Ice River have presumably survived because the plutonic mass was not penetratively redeformed or metamorphosed, but behaved as a competent mass in relatively less competent, enclosing early Palaeozoic sedimentary rocks.

The minimum age of Ice River Complex is indicated by the oldest K–Ar age so far obtained (360 m.y., Baadsgaard *et al.* 1961). Its maximum age is indicated by its intrusive relations with rocks of Lower Ordovician age (Allan, 1914, but see Gussow and Hunt, 1959, for contention of older age).

Mesozoic and Later Ages

Potassium–argon age determinations from southern British Columbia listed in Table I (Appendix) are plotted on Figure 2. Although the ages clearly spread throughout the Mesozoic, definite maxima appear to occur. Further, these maxima are similar to those on the accompanying plot of all other K–Ar measurements in the remaining part of British Columbia and Yukon Territory.

On Figure 4 an approximate outline is given of all the plutonic masses as well as the Shuswap Metamorphic Complex in this part of southern British Columbia. Also on the same diagram are shown the various maxima found on Figure 2. It is evident that no correlation of a single maximum with a single intrusion exists. It is further evident that an evaluation and understanding of the significance of the age groups cannot be attempted without a study of the occurrence of ages within single plutonic masses. The Adamant, White Creek, and Nelson batholiths are perhaps the most instructive plutons in this belt both with respect to variation in individual dates obtained, and from the information that can be extracted from this variation.

Adamant Batholith

The Adamant batholith occurs in the northern Selkirk Mountains, the northernmost pluton in a series of small intrusive masses along the Selkirk–Purcell Mountain arc (see Fig. 4). The petrology of this mass is currently being studied in detail as a Ph.D. thesis project by Mr. P. E. Fox at Carleton University; therefore conclusions expressed here are tentative pending the results of this study.

This pluton covers an area of about 60 square miles and has a long, oval shape with the major axis oriented east–west, at right angles to the structural trends in the surrounding region. Enclosing strata are wrapped conformably around the mass and an internal post-crystalline foliation around the periphery is also parallel to the contact of the pluton. Rock types found in the mass consist of biotite–hornblende granodiorite in the deformed outer zone of the pluton followed inward by a mafic-rich hornblende granodiorite. One large mass and two smaller masses of hypersthene monzonite are found in the interior of the pluton. Locally, foliation may be traced from hornblende granodiorite into hypersthene monzonite without divergence. The Adamant pluton has been emplaced into late Precambrian Horsethief Creek Group rocks (Wheeler 1961) of moderate grade of regional metamorphism. Late pegmatites and quartz-rich veins cut both monzonite and granodiorite.

Six K–Ar ages have been obtained from the Adamant mass (see Table I). Determinations on biotite from the biotite–hornblende granodiorite of the periphery yield 281 m.y. and 200 m.y. From the sample yielding 200 m.y. a hornblende separate gave 116 m.y. and from a similar rock, a potash feldspar gave 92 m.y. Biotite from a pegmatite and from a still later quartz-rich vein yielded ages of 131 m.y. and 90 m.y. respectively. Clearly the Adamant is not a simple post-tectonic intrusion.

At the present time, on the basis of available geological information as well as age determinations, the origin and emplacement of Adamant pluton may be considered somewhat as follows. The pluton had begun its evolution at depth some time before the late Palaeozoic and possibly at that time consisted largely of hypersthene monzonite. It was, at a later time, remobilized and reintruded as a competent, resistant, mass since the surrounding regionally metamorphosed sedimentary rocks are thrust aside and conformably envelop the pluton. Synchronous with, or prior to, re-emplacement of the pluton, mafic-rich hornblende granodiorite developed from hypersthene monzonite (P. E. Fox, pers. comm., 1963). This possibility is emphasized by the present continuity of foliation at some localities, through both granodiorite and monzonite. A post-crystalline foliation has developed, also presumably synchronously with re-emplacement, around the boundary of the pluton. It is from the deformed biotite–hornblende granodiorite near the periphery of the mass that the ages of 281 m.y. and 200 m.y. were obtained. They are, therefore, considered to be the minimum ages possible for the original consolidation of Adamant pluton.

Late horizontal swarms of pegmatite dykes cut both the granodiorites and the monzonite, and biotite from one such dyke yields an age of 131 m.y. It must, therefore, be assumed that final emplacement of Adamant batholith took place at or before this time. The latest events in this pluton are reflected in the ages of 116 m.y. on hornblende, and of 90 and 92 m.y. respectively from biotite in a late quartz-rich vein and potash feldspar in granodiorite.

Age determinations from the surrounding metamorphic terrain that extends some tens of miles north and south of Adamant pluton range from 73 m.y. to 205 m.y. (Lowdon 1963; Lowdon *et al.* 1963). Here also a long-continued succession of events is evident that extends from at least early Mesozoic and possibly earlier, to late Mesozoic (see Table I and Fig. 4). Deformation and metamorphism as well as the initial formation of Adamant pluton may well date back at least to late Palaeozoic. In this connection it is notable that two small masses of hypersthene monzonite in the western part of Adamant batholith are elongate in a northwestward direction and the associated foliation is also oriented northwestward parallel to the regional foliation in the surrounding rocks. This internal orientation may reflect the trend of early structures within the massif prior to emplacement in its present position. Therefore, it might be suggested that the northwesterly structural trends in this region were imposed much earlier than the late Mesozoic and may well have been very early Mesozoic or even Palaeozoic. Much further geological work must be done in this complex locality before such a suggestion can be checked.

Adamant batholith and its environs clearly demonstrate the inadvisability of selecting a single sample from even the smallest pluton or from metamorphic rocks and expecting an "absolute" result. Certainly any single age from Adamant batholith could have been accommodated with ease as a plausible "age" of emplacement of this mass; yet no clue would have been gained of its true complexity.

Toby Stock

This stock, on the Purcell divide at the head of Toby Creek (Fig. 4), consists of rocks of similar composition to those of Adamant batholith and range from hypersthene monzonite to mafic-rich granodiorite. It consists of a narrow, elongate body parallel to and conformable with the northerly trend of folds in the enclosing strata. It is about two miles wide at its northern part, but narrows to less than one mile at its southern limit. There is a marked, though narrow, contact aureole.

Two age determinations were obtained from this mass, one from a mildly foliated granodiorite from the wider northern part (T-IRA-1 at 232 m.y.) and one from the highly lineated southern limit (T-3RA-1 at 179 m.y.) There is continuous outcrop between these two localities and both samples are from the same rock unit. The only apparent difference consists in the strong development of a post-crystalline lineation in the narrow southern end of the mass that is perfectly parallel to the fold axes in the enclosing strata. Table III, of both modal and chemical analyses of these two rocks, emphasizes this similarity. There is no greater variation between the two specimens than may be obtained at either locality by analysing a number of samples.

TABLE III

Specimen	Modes, vol. %			Chemical analyses (rapid method)	
	T-1RA-1	T-3RA-1		T-1RA-1	T-3RA-1
Quartz	18.7	14.7	SiO_2	61.7	61.3
Plagioclase	34.4	45.8	Al_2O_3	16.9	16.3
Potash feld.	12.1	7.2	Fe_2O_3	2.0	2.2
Biotite	22.8	19.3	FeO	4.05	3.88
Hornblende	—	1.4	CaO	4.3	4.2
Epidote	10.2	10.9	MgO	2.3	2.1
Acc.	1.8	0.7	Na_2O	3.6	3.4
			K_2O	3.5	3.4
			H_2O	0.61	0.75
			TiO_2	0.70	0.89
			P_2O_5	0.21	0.31
			MnO	0.09	0.09
			CO_2	n.d.	0.04
Total	100.0	100.0	Total	100.0	98.9

A number of conclusions may be reached concerning Toby Stock.

(a) It is similar in composition to the Adamant mass, but occurs parallel to the trend of the major structures, not athwart them as with Adamant. Ages obtained, as with Adamant, are among the oldest (Mesozoic) in the belt. There is no evidence to indicate that Toby massif could not have been emplaced earlier than 232 m.y.

(b) It has undergone considerable deformation and recrystallization after consolidation, and has therefore been involved in tectonic events that clearly post-date its initial emplacement.

(c) The younger age obtained in the more deformed "tail" of the mass must be related specifically to the greater intensity of the post-crystalline deformation in this part of the stock. It is thus an approximate measure of the effectiveness of a structural episode alone in modifying an isotopic age. It is also possible that the structural event that caused the diversity of ages between the two parts of this pluton was not sufficiently intense to disperse fully all argon produced to that time. Thus both ages may well date no specific event, but may be partial remnants of an earlier age of emplacement of Toby massif. It must be noted, however, that either one or the other of these ages could have been given a plausible, regional, interpretation in terms of either emplacement time or tectonic event. They may be evaluated realistically only by considering in detail the geology of the pluton and its surroundings.

White Creek Batholith

White Creek batholith, in common with many other plutons in this belt, lies with its long axis across the regional trend of structures, in the enclosing Precambrian rocks (Reesor 1958). It consists of a succession of rock types

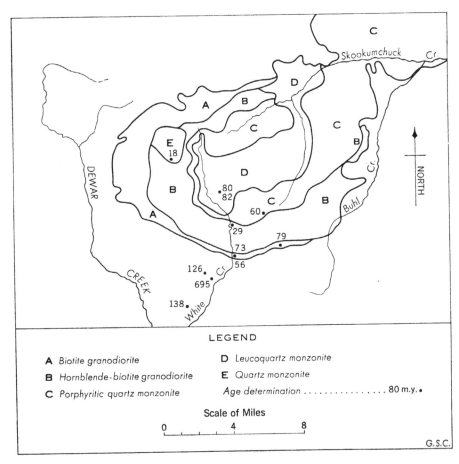

FIGURE 5. Zones in the White Creek batholith and location of age determinations.

ranging from biotite–hornblende granodiorite with megacrysts of potash feldspar and total mafic minerals ranging from 20–30 per cent, to leuco-cratic quartz monzonite with less than 5 per cent mafic minerals. The general arrangement of the principal rock varieties as well as the results of K–Ar age determinations are summarized in the accompanying Figure 5. Clearly the age of this pluton is not less than 76–82 m.y., the maximum ages obtained both from the outer border zone and from the late quartz–monzonite core respectively. However, a wide range of "ages" down to 18 m.y. is difficult to explain on the basis of available information.

In order to test the validity of these results a number of determinations by Rb–Sr on both individual minerals and whole rock samples have been carried out on each rock type in this pluton. These results will be published in detail when available. In general the Rb–Sr results tend to confirm the oldest ages obtained by K–Ar. Thus a specimen from the leucoquartz monzonite core yields, by Rb–Sr, an age of 84 m.y. on muscovite and 67 m.y. on coexisting biotite. On the other hand, biotite from two specimens collected near the periphery of the mass yield ages of 68 m.y. and 65 m.y. A whole-rock isochron (see for example, Hurley *et al.*, 1962, p. 109, for the underlying principle of the whole-rock isochron) for a series of specimens from the core of White Creek yields an age of about 90 m.y., and thus confirms the age of consolidation of the core of the batholith.

The results so far appear to yield a consistent age of consolidation with the oldest mineral ages on muscovite and biotite at 80–85 m.y. and Rb–Sr biotite ages at 65–68 m.y., with a series of K–Ar "ages" trailing down to 18 m.y. Perhaps this reflects minor recrystallization within the batholith or irregular cooling below a critical level at which argon can no longer be dissipated easily. No specific petrographic or other criteria suggest which explanation is correct or point to any other explanation for these low K–Ar ages. Nevertheless the Rb–Sr whole-rock isochron confirms one of the important maxima shown on Figure 2 at 90 m.y. Further such checks are planned in the near future for other age maxima shown in Figure 2 from other plutons in Southern British Columbia.

Nelson Batholith

Up to this point we have considered the small plutons in the eastern belt, and have found their emplacement, origin, and age patterns to be by no means simple. We may now consider Nelson batholith one of the two large plutons in the belt lying between these small intrusive plutonic masses and the Shuswap Metamorphic Complex (see Fig. 4 and 6).

The Nelson batholith is here considered to be the massive granitic pluton lying between Kootenay and Slocan Lakes, principally north of Kootenay Arm, though a narrow projection of the mass continues across Kootenay Arm as far south as Ymir Creek. (See, however, Little, 1960, pp. 81–82, for a more extensive definition of Nelson batholith.) In spite of the above limitation of the area of Nelson batholith this massif outcrops over 900 square miles. Rock types present vary from hornblende–biotite granodiorite, typically containing large megacrysts of potash feldspar, to biotite granodiorite, and leucoquartz monzonite. The hornblende-bearing granodiorite is dominant, but the more leucocratic varieties are found throughout the batholith not only as irregular masses of large extent, but also as dykelets and veinlets in the dark granodiorite. Along the west and southwest fringe of the batholith, hornblende syenodiorite and augite–hornblende monzonite occur as a later intrusion (see Fig. 6).

Structurally, much of this pluton is massive, though both vertical foliation and steep lineation may be found locally within it. The western

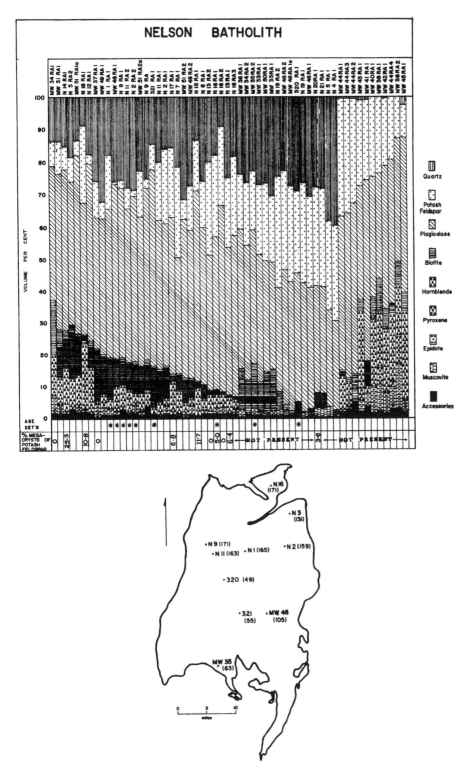

FIGURE 6. Compositional range of rocks in Nelson batholith and location of specimens for age determination. Number in brackets on sketch map is K–Ar age in million years.

boundary, along Slocan Lake and southward, is marked by a profound break that has been a locus of deformation and later intrusion over a long period (Reesor 1963).

Crosby (1960) describes what he considers a protoclasis or deformation and cataclasis synchronous with the emplacement of the pluton along part of the eastern edge of Nelson batholith north of Kootenay Arm. In 1962 one of us examined the southern "tail" of the massif outcropping along Ymir Creek. The rocks there show strong post-crystalline deformation, and an intense lineation parallel to the fold axes and cleavage-bedding intersection lineation in the enclosing rocks. Thus the narrow southern "tail" of Nelson batholith has been involved in at least the latest period of intense structural deformation that affected the enclosing sedimentary and metamorphic rocks.

Eleven K–Ar age determinations are available from localities throughout the main mass of Nelson batholith. The "ages" fall into three groups. (Table I [Appendix] and Fig. 6). The older group consists of ages reported here from the massive hornblende–biotite granodiorite. Two samples, one from a point a few miles east of the south end of Slocan Lake, and one from the northernmost, nearly separate mass of Nelson batholith, yield ages of 171 m.y. Three further specimens, taken in a rough east–west line across the central part of the pluton respectively near Arlington Peak (163 m.y.), Nansen Mt. (165 m.y.), and Pontiac Peak (159 m.y.) yield similar ages.

The younger group is represented by two specimens previously reported, 320 RA-1 (60-21 in Lowdon 1961) from a leucoquartz monzonite at the head of Duhamel Creek and 321 RA-1 (60-22), from a hornblende–granodiorite about nine miles downstream from the above specimen, and yield respectively 49 m.y. and 55 m.y. A further, young age is reported here; specimen MW35RA-2, a biotite granodiorite, 2½ miles west of the town of Nelson, yields an age of 63 m.y.

Intermediate ages between these two extremes range from 86 m.y. to 131 m.y. Specimen N-3RA-1 (61-17 in Lowdon et al. 1963) from Mt. Chipman in the northeastern part of the pluton at 131 m.y. is the oldest. Specimen MW48RA-1 of hornblende–biotite granodiorite was collected on the main highway about 11 miles east of Nelson, and yields an age of 105 m.y. Finally an age of 86 m.y. is reported by Little (1960, pp. 86–87) from a specimen collected west of Nelson.

Although much work remains to be done on this structurally and petrogenetically complex mass, yet a tentative suggestion as to the broad outlines of its formation and emplacement seems warranted at this time. This is particularly so since structural relations and K–Ar dates within this massif are not apparently contradictory. It may be suggested that Nelson batholith developed in three main stages.

1. Emplacement and consolidation of a large massif of hornblende–biotite granodiorite before 171 m.y., at greater structural depth than it now

exists. These ages must be considered a minimum and this episode may well be considerably older. The older ages have been preserved only because the early granodiorite reacted to later structural events as a competent mass, and was not penetratively redeformed.

2. Mobilization and reintrusion of a relatively passive mass resultant upon an intense structural episode that affected the region generally. This might have taken place at, or later than, 131 m.y., or latest Jurassic – earliest Cretaceous. This type of emplacement would imply a profound effect on the structural patterns of the surrounding region, and may, in fact, be the fundamental cause of the development of the Kootenay Arc.

3. Further emplacement of leucocratic quartz monzonite in and around the fringe of the batholith, and still later emplacement of augite–hornblende monzonite and syenodiorite west and southwest of the main batholithic mass.

It is proposed to continue structural and petrologic studies, as well as to make further isotopic age determinations by Rb–Sr, to test or to discard the above tentative hypothesis of the evolution of Nelson batholith.

In some localities, in the southern part of Nelson batholith, granitic rocks are intrusive into rocks of lower Jurassic age (Toarcian, Little, 1960, p. 67, and Frebold and Little, 1962) and thus these plutons must be much younger than the ages indicated from the granodiorite in the main mass if Kulp's time-scale is accepted. Yet, if the views expressed above are correct, there is no conflict betwen the age data, stratigraphic data, and structural data. Nelson batholith may have evolved at depth during the time of formation of Rossland Formation or even earlier, during the Triassic.

In this context it is interesting to consider the results of age measurements on Guichon batholith (see Table I and Fig. 4). Baadsgaard *et al.* (1961, p. 696) report an age of 186 m.y. from this pluton that apparently agrees well with the lower Jurassic age of the mass established on stratigraphic grounds by Duffell and McTaggart (1952, p. 79). Additional measurements from a number of different samples within the central part of this mass yield an average age of 235 m.y., approximately the base of the Mesozoic on Kulp's time-scale. Pending further determinations by the K–Ar method on specimens collected from the periphery of the pluton, it may be suggested that its history is somewhat similar to that assumed for Nelson batholith. Thus an older pluton has been moved with additional peripheral intrusion such that stratigraphic relations were established by the latest major event in a long chain of earlier events leading to the formation of Guichon batholith. Further work is currently being carried out on specimens collected near the boundary of this mass, and it is hoped that a whole-rock Rb–Sr isochron can be done in the near future.

Throughout the examples presented so far there is evidence not only from the age determinations but also from extensive structural and petrologic information that individual masses have evolved through a very long time, in some cases during the entire Mesozoic. Complex masses have been presented as examples in the above discussions, partly because these are the ones studied in most detail and partly because they are the most significant in the long-range attempt to learn something of the plutonism in a mountain belt.

It might be deduced that further study on seemingly simple masses will reveal unsuspected complexities. For example, Bugaboo batholith has a fine-grained granodioritic western segment cut by younger quartz monzonite of the main mass. Yet, the geologically younger quartz monzonite yields an age of 130 m.y. versus 100 m.y. for the granodiorite (see Table I and Fig. 4). Horsethief batholith yields an age of 200 m.y. from its western fringe and 108 m.y. from its main eastern part. It too must have a complex origin; yet a cursory reconnaissance shows a relatively small, fairly uniform mass, and the rock yielding the older age is a massive biotite granite not unlike those of much younger age obtained in the general region.

Bayonne batholith has a range of composition from quartz diorite to leucoquartz monzonite. It's structural features vary from conformable with the enclosing metamorphic rocks, to sharply cross-cutting and intrusive. Ages obtained range from 33 m.y. to 100 m.y. (Table I) and these results taken together with other age patterns on similar rocks in the region serve to bring out the probable complexity of origin of a small granitic pluton coloured pink on most geological maps.

It is evident that K–Ar dates are particularly suitable, especially in an initial reconnaissance for identifying hitherto unsuspected complexities. It is obvious, however, that a few scattered K–Ar determinations must be considered as somewhat less than "absolute" in their application to specific petrogenetic or tectonic events. It is necessary to consider many ages along with all available petrologic, stratigraphic, and structural evidence.

The Shuswap Metamorphic Complex

The individual plutonic masses discussed above have been moved from the environment of their ultimate origin and, however much further evolution they have undergone, either synchronous with their removal or after they have come to rest, they have been separated from their associated rocks in the original environment. Within the region of the Shuswap Metamorphic Complex, however, a large part of the evolution, both structural and petrogenic, has apparently taken place *in situ*. In consequence a well-established, long succession of events is recorded in both the structural and metamorphic-plutonic history of this complex, deep, central zone of the mountain belt in southern British Columbia.

Table IV summarizes briefly the parallel development of a complex structural and plutonic history in Valhalla Range along the southeastern limit of the Shuswap immediately west of Nelson batholith (see Fig. 4). (Summarized from Reesor, 1963.)

Although there is here, and throughout the Shuswap, much evidence of

TABLE IV

SUMMARY OF EVENTS IN VALHALLA COMPLEX

		Structural evolution	Petrogenic evolution
Late stage	Locally confined	8. Late faulting in Crushed Zone 7. Minor faulting and jointing in complex 6. Igneous emplacement under regional tensional conditions	8. Pyritization 7. Basalt dyke intrusion in western Valhalla. Sericitization and epidotization in Crushed Zone east of Valhalla Complex 6. Intrusion of hornblende and/or augite monzonite both east and west of Valhalla Complex
Second phase	Destructive	5. Diapiric upward movement of cores of Valhalla and Passmore Domes 4. Strong, penetrative, cataclastic flowage in all layers of all rock types in eastern part of Valhalla Complex. Easiest relief of stress vertically upward	5. Possibly some synchronous emplacement of massive leucogranite in zone west of Valhalla Complex 4. Minor pegmatite and pegmatitic granite. Metamorphism in garnet–sillimanite subfacies of amphibolite facies
First phase	Constructive	TERMINATION OF PRINCIPAL CONSTRUCTIONAL PHASE OF EVOLUTION	
First phase	Constructive	3. Minor structural impulses, with gradual growth of antiformal masses. Shape and location of Passmore Dome dependent on formation of Valhalla Dome and thus slightly later 2. Initial uplift of Valhalla Dome with original form determined as an east-west elongate antiformal complex. Coincident flowage of Hybrid Gneiss off flanks of Valhalla Dome governed by shape of confining space. Initial syntectonic movement confined such that easiest relief of stress horizontal	3. Leucogranite and early pegmatite in all layers of the complex. Older grey leucogranodiorite gneiss in all layers 2. Metamorphism to perhaps hornblende granulite facies, with partial melting of pelitic and semipelitic rocks. Horizontal emplacement of quartz diorite and granodiorite of Veined Gneiss in Valhalla Dome and upper layer (Mixed Gneiss) of Passmore Dome
First phase	Constructive	1. Possible earlier structural episodes not recognized	1. Deposition of thick succession of pelites, semipelites, quartzite, and minor limestone. Minor conformable basic igneous rocks

superimposition of a number of events over a long period (Jones 1959; Wheeler, pers. comm.), the age determinations from many localities and many different local environments throughout the Shuswap yield ages of Cretaceous, or younger, by K–Ar determinations on either biotite or muscovite (Table I [Appendix] and Fig. 4). The key to this uniformity of results lies in the character of the last main phase of deformation as shown in Valhalla Complex. In some localities a penetrative redeformation has occurred in a zone of sillimanite–almandine grade of regional metamorphism. Even in localities in which no visible deformation in the second phase is evident, quartz microfabric maxima show similar patterns, through more diffuse concentrations than those in highly deformed zones. Thus, even

massive rocks of granitic composition appear to have been affected penetratively by the later deformation. Since both biotite and quartz have been
rearranged visibly in part of Valhalla Complex, we must conclude from the
age results that the biotite has also been affected in massive zones not visibly
affected by the second main phase of deformation.

Ages from Valhalla Complex range from 66 m.y. to 11 m.y. (Reesor, *in*
Lowdon 1961) and from the remaining part of the Shuswap from 102 m.y.
to 36 m.y. (See 31 determinations listed in Table I.) These isotopic determinations must, everywhere in the Shuswap, represent ages no older than
the last main phase of penetrative redeformation. This is an important
episode of tectonism in the Shuswap for it was at this stage, for the first
time, that the northerly striking Cordilleran structural trends were superimposed upon the dominantly east–west earlier structures of the metamorphic complex in the core of the mountain belt. As in Valhalla (see
Table IV), there are later events at scattered localities throughout the
Shuswap, minor intrusions, faults and shears, emplacement of basaltic dykes,
and even emplacement of minor leucogranite. Ages younger than 80–100
m.y. may in some cases reflect these younger events or may reflect irregular
unroofing and cooling. Many of the very young ages obtained must be
anomalous, for Tertiary rocks of Oligocene (?) age are found unconformably overlying Shuswap metamorphic rocks along their western outcrop limit (Jones 1959, p. 52). Measurements by the K–Ar method on
similar Tertiary rocks a little farther west yield ages of 45–50 m.y. (Rouse
and Mathews 1961) and therefore ages younger than these in the high-grade
metamorphic rocks do not represent a penetrative, wide-spread structural
or metamorphic event. Furthermore, a pegmatite that cuts across the
metamorphic rocks and the latest structures yields an age of 81 m.y. (Spec.
CO–61–187, Table I). Therefore, the latest penetrative structural and
metamorphic event in the Shuswap cannot be younger than 81 m.y. and
must date back to 102 m.y. or older.

Another anomalous feature of the Shuswap age determinations is the
96 m.y. age reported by Baadsgaard *et al.* (1961) for the massive granite
rock of Monashee Pass. These massive quartz monzonites penetrate and
engulf the rocks of the metamorphic terrain. Although their emplacement
may be broadly synchronous with the last major stage of deformation in the
Shuswap, it is certain that the metamorphic terrain does not post-date these
massive granitic rocks.

The K–Ar ages obtained from the Shuswap to date confirm the penetrative nature of the latest major structural event in the Shuswap during
the imposition of northerly Cordilleran trends, at high temperature and
pressure, upon the ductile core of this part of the mountain belt. In spite
of great variation in rock types present, this terrain has reacted as an
essentially homogenous mass and virtually no single grain has remained
without the imprint of this deformation. Further, the dates so far available
broadly confirm the synchroneity of this major event in the Shuswap with

the emplacement of some of the plutonic masses east of the core of the mountain belt.

It is evident from the structural patterns and from the ages so far obtained that the K–Ar method will nowhere penetrate the older events recorded in the structural fabric of the Shuswap Metamorphic Complex. As a first step in penetrating the latest event a whole-rock Rb–Sr isochron is now being obtained from a granodiorite augen gneiss in the core of Valhalla gneiss dome. This, if successful, should date the initial event in the early formation of east–west structures in Valhalla Complex, and will perhaps serve to relate this major tectonic event with plutonic and structural events in the zones of greatly contrasting structural and metamorphic fabric to the east of the Shuswap.

CONSIDERATION OF K–AR AGE RESULTS

Before briefly summarizing the results and the applicability of K–Ar determinations it is as well to consider very briefly the precision and replicability of the K–Ar measurements. Only in White Creek have measurements by the K–Ar method been checked by the Rb–Sr. There is broad agreement in measurements by Rb–Sr on individual minerals and on the age of the core of the massif by a whole-rock isochron (see discussion of White Creek above).

In some cases separate argon extractions from the same specimen have been checked, if there was reason to believe results to be incorrect (e.g. Guichon batholith in which stratigraphic relations appear to indicate a younger age than is possible from the K–Ar ages). In the case of Guichon, measurements at different times with different spike material on different specimens all yield similar results and one sample checked with a second argon extraction resulted in a difference of less than 3 per cent.

In general, a total error of 8 per cent is the maximum considered likely for measurements in this range of ages. Thus we must consider that the resulting ages, even within single plutons, represent significant events and are not errors of measurement. The essential problem is resolved into an estimate of which ages represent which specific events and which are hybrids of successive phases of metamorphism, structural deformation, or quiescent cooling within a single tectonic cycle. By and large we must accept the fact that the precision of K–Ar determinations is far superior to the accuracy or applicability of the interpretations of the results obtained. Meaningful interpretations may only be made if there is a great body of corroborating evidence.

The principal accomplishments of the K–Ar work may be summarized under three broad headings.

1. K–Ar ages have served to identify plutonic and metamorphic events belonging to earlier cycles of mountain-building even though the localities in which such ages have been found have been involved in the culminating

Mesozoic orogeny of the Cordillera. (See Precambrian ages and Ice River complex in the foregoing discussions.)

2. K–Ar ages confirm and emphasize the complexity of plutonism in the Mesozoic cycle of mountain-building. Even though individual plutonic masses have been subjected to a succession of events that extended throughout much of the Mesozoic, K–Ar ages have been preserved that broadly confirm the petrologic and structural, as well as stratigraphic evidence. (See, for example, Adamant, Nelson, and Cassiar batholiths in the foregoing discussions.)

3. K–Ar ages point to, and emphasize, a time relationship between plutonic masses or plutonic events not otherwise evident from structural or stratigraphic information available at the present time. One of the most interesting examples of this is the demonstration of the broad synchroneity of the last penetrative event in the Shuswap Metamorphic Complex and the emplacement of leucocratic granitic rocks in the lesser metamorphosed fringing zones of the Selkirk–Purcell mountain arc. Perhaps other methods of isotopic age determinations (zircon or whole-rock Rb–Sr) will serve to show the relationship, if any, between earlier events in the Shuswap and those plutonic masses to the east that show not only older K–Ar ages but appear also to be older on structural and petrogenic grounds. (See, for example, Adamant and Toby massifs and Nelson and Guichon batholiths).

From these results the conclusion seems fully warranted that K–Ar ages, providing they are obtained in sufficient numbers, are particularly suited to a broad reconnaissance timing of both plutonic and structural events in a mobile belt. Timing by other methods of isotopic age determinations coupled with much associated geological information must be added before we can advance beyond the bare reconnaissance.

The principal limitations in interpretation of K–Ar results may be summarized in the light of these discussions.

1. Effect of superimposition of two entirely different cycles. Clearly, in this case the effect of superimposing a Mesozoic on a Precambrian cycle has resulted in a trail of K–Ar ages ranging from 700–800 m.y. to less than 150 m.y. Thus no reason exists to conclude that 700–800 m.y. represents anything more than a minimum for the Precambrian metamorphism and plutonism in the southern Purcell and Rocky Mountains. Perhaps this event can be dated more specifically by whole-rock Rb–Sr, certainly not by either K–Ar or Rb–Sr on individual, separated, minerals.

2. Effect of tectonic and plutonic evolution in a single cycle (Mesozoic in this case). The above discussions have made it abundantly clear that the interpretation of K–Ar ages cannot be based upon the simple concept that plutonic rocks are simply emplaced during a mountain-building cycle and preserve no remnants of earlier development and no effect of later events in the tectonic evolution of the mobile belt. Quite the contrary, we have shown the involvement of many plutons in successive events throughout the entire Mesozoic.

A wide range of numbers is available together with a long-continued, though pulsatory, tectonic evolution in the mobile belt. Thus many plausible pigeon-holes exist for an age measurement to be fitted into. As yet, no specific event may be related with certainty to any specific measurement from a single pluton. Certainly one or two determinations from Adamant or Toby plutons would not have indicated their true complexity; yet any one of these determinations could have been given a plausible, though incorrect, interpretation. Further, any individual age may be a compound of many elements, a plutonic event, a structural event, or simply loss of daughter isotope during quiescent cooling, or any combination or superimposition of any number of such individual events. No specific method exists for identifying the precise degree to which any one or any number of the above elements has contributed to a particular isotopic ratio that is measured as an age. Thus specific interpretation of a few age determinations must be made with caution and only if much associated, contributory evidence is available.

Concluding Statement

The development of plutonic rocks has been presented briefly in the foregoing discussions from a region in northern British Columbia in which Mesozoic stratigraphy has been available to establish a broad pattern of plutonic emplacement, to a region in southern British Columbia in which interrelationships are dependent more on structural patterns than on detailed stratigraphic information. The latter area shows the development of plutonic rocks in belts of greatly contrasting structural style and regional grade of metamorphism.

Broadly, in northern British Columbia stratigraphic and structural evidence, combined with K–Ar age determinations, suggests a granitic succession from quartz diorite in early Mesozoic to miarolitic leucogranite in early Tertiary. In southern British Columbia there is broad confirmation of a succession of early hypersthene monzonite (that may date from the Palaeozoic?) to a quartz–diorite leucogranite succession in the Mesozoic, terminating with a syenodiorite, monzonite, leuco-syenite, and leucoquartz monzonite succession in the Tertiary.

Therefore, to date, it is possible to show, on the basis of both K–Ar and associated geological studies, the general evolution of plutonic rocks from the late Palaeozoic (?) to the Tertiary during the development of the principal orogenic cycle in these two areas of the Canadian Cordillera. No attempt is made here to relate particular episodes in this plutonic evolution to particular "orogenies." The plutonic evolution evident so far, from this study, lends a unity to the entire process of tectonic development in a mountain belt. There appears to be a succession of pulsatory events at about 30 m.y. intervals, established in part by stratigraphy, and in part by K–Ar age determinations. This evolution does not accommodate a specific "orogeny" whether Nevadan or Laramide (Coast Range or Rocky Mountain orogenies of White, 1959). On the contrary associated tectonic and plutonic evolution throughout a period embracing a much longer time shows a greater unity than is evidenced in the term Nevadan or Laramide as separate orogenic events.

There exists a clear pattern of large numbers of ages in the general range from 100 m.y. to 60 m.y. (Cretaceous) commonly though not invariably from leucocratic granitic rocks. This may reflect truly an increasing crescendo of plutonism during the progressive development of the tectonic cycle. It could, however, at least in part, reflect the successive destruction of older phases of the granitic succession by younger, or the structural reworking of older plutons by successively younger tectonic events. This possibility is certainly not contradicted by the evidence presented here for long-continued plutonic and associated tectonic evolution in a mountain belt.

It is apparent, furthermore, that a simple granitic sequence in time ranging from autochthonous gneiss and migmatic to allochthonous migma and intrusive granite mobilized only in the dying stages of orogenic activities is not confirmed by the results so far available in southern British Columbia. Even some of the smallest plutons have evolved, in successive pulses, over a very long time. There is also evidence that the evolution of a gneissic terrain (Shuswap) continues throughout the orogenic cycle, though the character of the metamorphism and structural style may change at intervals dependent upon successive, pulsatory changes in temperature and pressure conditions in the core of the mountain belt.

Future investigation of these problems must involve further reconnaissance by K–Ar in northern British Columbia, to establish better some of the details of the development of granitic and metamorphic rocks. Investigation in southern British Columbia must involve further penetration of early events in the plutonic cycle by Rb–Sr whole-rock or by zircon studies coupled with further detailed geological work.

References

ALLAN, J. A. (1914). Geology of field map-area, British Columbia and Alberta. Geol. Surv. Can., Mem. 55.

BAADSGAARD, H., FOLINSBEE, R. E., and LIPSON, J. (1961). Potassium–argon dates of biotites from Cordilleran granites. Bull. Geol. Soc. Am., 72: 689–701.

CROSBY, P. (1960). Structure and petrology of the Central Kootenay Lake area, British Columbia. Unpublished Ph.D. Thesis, Harvard University.

DUFFELL, S. and McTAGGART, K. C. (1952). Ashcroft map-area, British Columbia. Geol. Surv. Can., Mem. 262.

FREBOLD, H. and LITTLE, H. W. (1962). Palaeontology, stratigraphy, and structure of the Jurassic rocks in Salmo map-area, British Columbia. Geol. Surv. Can., Bull. 81.

FYLES, J. T. (1962). Two phases of deformation in the Kootenay arc. Western Miner and Oil Rev., 35 (no. 7): 20–26.

GABRIELSE, H. (1962a). Cry Lake map-area, British Columbia. Geol. Surv. Can., Map 29-1962.

——— (1962b). Kechika map-area, British Columbia. Geol. Surv. Can., Map 42-1962.

—— (1963). McDame map-area, Cassiar District, British Columbia. Geol. Surv. Can., Mem. 319.

GABRIELSE, H. and SOUTHER, J. G. (1962). Dease Lake map-area, British Columbia. Geol. Surv. Can., Map 21-1962.

GABRIELSE, H. and WHEELER, J. O. (1961). Tectonic framework of southern Yukon and northwestern British Columbia. Geol. Surv. Can., Paper 60-24.

GUSSOW, W. C. and HUNT, C. W. (1959). Age of Ice River Complex, Yoho National Park, British Columbia. J. Alta. Soc. Petrol. Geol., 7: 62.

HUNT, G. (1962). Time of Purcell eruption in southeastern British Columbia and southwestern Alberta. J. Alta. Soc. Petrol. Geol., 10: 438–42.

HURLEY, P. M., FAIRBAIRN, H. W., FAURE, G., and PINSON, W. H. JR. (1962). New approaches to geochronology by strontium isotope variations in whole rocks. Dept. of Geol. and Geophys., Mass. Inst. Technol.; U.S. Atomic Energy Comm., Tenth Annual Progr. Rep. NYO-3943, pp. 109–13.

JONES, A. G. (1959). Vernon map-area, British Columbia. Geol. Surv. Can., Mem. 296.

KULP, J. L. (1961). Geologic time scale. Science, 133: 1105–14.

LEECH, G. B. (1962). Metamorphism and granitic intrusions of Precambrian age in southeastern British Columbia. Geol. Surv. Can., Paper 62-13.

LITTLE, H. W. (1960). Nelson map-area, east half. Geol. Surv. Can., Mem. 308.

LORD, C. S. (1948). McConnell Creek map-area, Cassiar District, British Columbia. Geol. Surv. Can., Mem. 251.

LOWDON, J. A. (1960). Age determinations by the Geological Survey of Canada, Rep. 1 —Isotopic ages. Geol. Surv. Can., Paper 60-17.

—— (1961). Age determinations by the Geological Survey of Canada, Rep. 2— Isotopic ages. Geol. Surv. Can., Paper 61-17.

—— (1963). Age determinations by the Geological Survey of Canada, Rep. 3— Isotopic ages. Geol. Surv. Can., Paper 63-17.

LOWDON, J. A., STOCKWELL, C. H., TIPPER, H. W., and WANLESS, R. K. (1963). Age determinations and geological studies (including isotopic ages—Rep. 3). Geol. Surv. Can., Paper 62-17.

MULLIGAN, R. (1955). Teslin map-area, Yukon Territory. Geol. Surv. Can., Paper 54-20.

POOLE, W. H. (1956). Geology of the Cassiar Mountains in the vicinity of the Yukon – British Columbia Boundary. Princeton University, Ph.D. thesis.

POOLE, W. H., RODDICK, J. A., and GREEN, L. H. (1960). Wolf Lake map-area. Geol. Surv. Can., Map 10–1960.

REESOR, J. E. (1958). Dewar Creek map-area, British Columbia. Geol. Surv. Can., Mem. 292.

—— (In press). Structural evolution and plutonism in Valhalla Complex. Geol. Surv. Can., Bull.

ROUSE, G. E. and MATHEWS, W. H. (1961). Radioactive dating of tertiary plant-bearing deposits. Science, 133: 1079–80.

ROOTS, E. F. (1954). Geology and mineral deposits of Aiken Lake map-area, British Columbia. Geol. Surv. Can., Mem. 274.

WATSON, K. de P. and MATHEWS, W. H. (1944). The Tuya Teslin area, Northern British Columbia. B.C. Dept. Mines, Bull. 19.

WHEELER, J. O. (1961). Rogers Pass map-area, British Columbia. Geol. Surv. Can., Prel. Map 4-1961.

WHITE, W. H. (1959). Cordilleran tectonics in British Columbia. Bull. Am. Assoc. Petrol. Geologists, 43: 60–100.

[The Appendix follows on pp. 130–138.]

APPENDIX

TABLE I

List of K–Ar Ages

(M after age signifies K–Ar determination on muscovite; otherwise all determinations on biotite. Physical data for all GSC determinations are published in Lowdon 1960, 1961, 1963, and Lowdon et al. 1963)

Sample number	Reference number	Location	Rock type	Age
XXVII	GSC 61–42	60°00'30"N., 132°08'20"W.; Mile 777.7 Alaska Hwy., Yukon Territory	Quartz–muscovite schist	222(M)
Composite sample	GSC 59–9	Same general area as above extending northwest from Mile 777.7	Mica schist	214(M)
VIII-M-59	GSC 62–73	59°57'N., 131°58'W.; Mile 764 Alaska Hwy., British Columbia	Chlorite–mica–albite–quartz gneiss	194(M)
GA 21/9/61/4	GSC 62–70	58°58'36"N., 130°01'24"W.; west of Dease River	Mica–quartz schist	178(M)
VIII-M-59	GSC 62–72	59°57'N., 131°58'W.; Mile 764 Alaska Hwy., British Columbia	Chlorite–mica–albite–quartz gneiss	124
GC 59–218a	GSC 60–30	60°37'N., 130°47'W.; 3 miles from Marker Lake batholith	Quartz–feldspar biotite schist	98
GA 22/9/61/3	GSC 62–74	59°30'40"N., 128°55'45"W.; north end of Horseranch Range	Biotite–muscovite–quartz–feldspar gneiss	57(M)
GA 2/9/61/2A	GSC 62–71	58°08'30"N., 129°52'00"W.; Hotailuh batholith	Biotite–hornblende granodiorite	193
GAC 17/8/61/A₂	GSC 62–68	59°27'N., 127°42'W.; Cassiar batholith near Dall Lake	Muscovite–biotite quartz monzonite	139(M)
GC 59–218	GSC 61–45	60°33'N., 130°57'W.; Marker Lake batholith	Granodiorite	126
GAC 17/8/61/A₂	GSC 62–69	59°27'N., 127°42'W.; Cassiar batholith near Dall Lake	Muscovite–biotite quartz monzonite	123
Baadsgaard, et al. (1961)	AK 50	60°04'N., 130°29'W.; Cassiar batholith, Alaska Hwy.	Quartz monzonite	101
Rd-59-93	GSC 60–28	60°32'N., 131°29'W.; north end Cassiar batholith southeast of Wolf Lake	Biotite–granodiorite	98
GA 8/9/59/1	GSC 60–25	58°48'18"N., 130°01'00" W.; Cassiar batholith near Dease River	Quartz monzonite	71
5-1-2/P	GSC 59–14	60°02'32"N., 131°10'11"W.; Seagull batholith	Leucoquartz monzonite, locally miarolitic	59

TABLE I (Continued)

Sample number	Reference number	Location	Rock type	Age
		PRECAMBRIAN AGES (SOUTHERN PURCELL MOUNTAINS)		
LD-ML-7	62–41	2 miles west of Alki Creek; 49°37′24″N., 116°15′85″W.		555(M)
	59–3	Sullivan Mine; 49°43′N., 116°00′W.	Lamprophyre	580
I 525-5-2	61–25	Angus Creek, 0.9 mile above Hellroaring Creek; 49°34′03″N., 116°09′06″W.	Pegmatitic granodiorite	675(M)
W29-RA-1	62–1	In large sill 1.88 miles south of White Creek batholith and west of the Creek; 49°47′18″N., 116°18′54″W.	Pegmatite with very coarse books of muscovite	695(M)
	60–2	Hellroaring Creek 1.8 miles south of St. Mary Lake; 49°34′30″N., 116°10′42″W.	Granodiorite of varied texture partly coarse to pegmatitic with tourmaline and beryl	705(M)
LD-ML-1	61–26	West of Matthew Creek and 5¾ miles southwest of Kimberley; 49°38′48″N., 116°05′40″W.	Pegmatitic leucocratic granodiorite	745(M)
	59–2	Sullivan Mine; 49°43′N., 116°00′W.	Lamprophyre	765
LD-ML-3	61–27	West of Matthew Creek and 5½ miles southwest of Kimberley; 49°38′48″N., 116°05′17″W.	Quartzite	790(M)
		ICE RIVER COMPLEX		
Baadsgaard et al.	AK 84	51°11′N., 116°27′W.	Pegmatite	304
Ice 4-RA-1A	59–8	51°12′N., 116°29′W.	Pyroxene-rich melanocratic rock – coarse biotite, commonly along blades of pyroxene	330
Ice 5-RA-1	59–7	51°12′N., 116°29′W.	Syenite dyke with coarse tablets of biotite	340
Baadsgaard et al.	AK 46	51°10′N., 116°23′W.	Jacupirangite	355
Baadsgaard et al.	AK 83	61°08′N., 116°24′W.	Minette	360
		ADAMANT BATHOLITH		
FK-A-4	61–24	Rock bastion on north side of Granite Glacier; 51°46′08″N., 117°54′20″W.	Quartz-rich vein in granodiorite	90
5FK-7-2	62–24	North side of Wotan Glacier 1 mile northeast of Wotan Peak; 51°44′30″N., 118°44′00″W.	(Potash feldspar)	92
FK-A-2	61–22	Elevation 7000 feet on north side of Palmer Creek; 51°42′40″N., 117°50′30″W.	Pegmatite	131

TABLE I (*Continued*)

Sample number	Reference number	Location	Rock type	Age
FK-A-3	61–23	Elevation 6500 feet south of tip of Granite Glacier; 51°46'08"N., 117°51'30"W.	Biotite–hornblende granodiorite	200
	62–25		Hornblende from above granodiorite	116
FK-A-1	61–21	Elevation 7000 feet on north side of Palmer Creek; 51°42'40"N., 117°50'30"W.	Biotite–hornblende granodiorite	281
METAMORPHIC ROCKS NORTH AND SOUTH OF ADAMANT BATHOLITH				
WB-120-1-61	62–49	6 miles S. 81°W. of mouth of Swan Creek; 51°54'10"N., 117°56'08"W.	Coarse-grained mica schist	72(M) 73
WB-120W-3-61	61–28	6.5 miles S. 84°W. of mouth of Swan Creek; 51°48'N., 117°57'W.	Pegmatite that cuts foliation in metamorphic rocks of Horsethief Creek group	107(M)
WB-121-1-61	62–51	0.5 mile S. 34°W. of Six Mile Pass; 51°34'55"N., 117°38'4"W.	Medium-grained mica schist	119
	62–52		Same	124(M)
WB-151-AD-61	62–53 62–54	0.65 mile N. 69°E. of Cupola Mountain; 51°34'05"N., 117°33'30"W.	Faintly crenulated mica schist	146 205(M)
FANG CREEK STOCK				
WB-129-W	62–23	West side of Tangier River; 51°20'N., 117°50'W.	Hornblende–biotite quartz monzonite with 14.9 per cent megacrysts of potash feldspar	168
BATTLE BATHOLITH				
WB-137-W-7	62–21	Freeze Cirque in Battle Range	Fine- to medium-grained quartz monzonite	91(M)
WB-137-W-6	62–20	Same	Same	92
	62–19	Freeze Cirque	Muscovite–biotite leucoquartz monzonite	94
WB-137-W-8	62–22	Freeze Cirque	Pegmatite	120(M)
BUGABOO BATHOLITH				
B-1-RA-1	61–20	On east side of glacier at head of first north branch of East Creek; 50°44'6"N., 116°55'30"W.	Biotite granodiorite	100
B-2-RA-1	62–17 62–18	On west side of upper part of glacier east-northeast of Howser Spires; 50;45'36"N., 116°49'06"W.	Leucoquartz monzonite Same	132 138(M)

TABLE I (*Continued*)

Sample number	Reference number	Location	Rock type	Age
		HORSETHIEF CREEK BATHOLITH		
H-2-RA-1	62–16	Elevation 8400 feet at lake in saddle on divide between Stockdale, Horsethief, and Forster Creeks; 50°36′12″N., 116°30′42″W.	Porphyritic quartz monzonite	108
H-1-RA-1	61–19	South of head of Howser River at 8200 feet; 50°38′6″N., 116°35′42″W.	Coarse biotite granite	205
		GLACIER CREEK STOCK		
G-1-RA-1	61–18	At south end of glacier that caps the stock; 50°23′12″N., 116°47′49″W.	Biotite granodiorite	127
G-1-RA-2	62–15	300 feet from above location; 50°23′12″N., 116°47′42″W.	Biotite granodiorite	145
		TOBY STOCK		
T-3-RA-1	62–14	Elevation 7750 feet in "tail" of stock; 50°11′N., 116°34′W.	Hornblende–biotite auger granodiorite gneiss	179
T-1-RA-1	62–13	Elevation 8600 feet at south end of Toby Glacier; 50°12′36″N., 116°33′24″W.	Biotite granodiorite	232
		FRY CREEK BATHOLITH		
1955-165a	60–18 60–19	3 miles up Fry Creek from footbridge crossing; 50°05′N., 116°51′W.	Leucoquartz monzonite	45 63(M)
F-8-RA-1	62–12	Elevation 7800 feet in "tail" of batholith east of head of St. Mary River; 49°52′N., 116°34′W.	Leucogranodiorite	76
F-1-RA-1	62–8	Cirque below Mt. Tyrrel; 49°59′00″N., 116°44′12″W.	Leucoquartz monzonite showing some evidence of late crystalline shearing	83(M)
F-3-RA-1	62–11 62–10	Southeast shoulder of Mt. Pambrun; 50°05′24″N., 116°33′24″W.	Fresh leucoquartz monzonite	86 91(M)
F-2-RA-1	62–9	Elevation 7250 feet southeast of junction of Fry Creek and Gillis Creek; 50°01′54″N., 116°39′42″W.	Fresh leucoquartz monzonite	97(M)
		WHITE CREEK BATHOLITH		
W-58-RA-21	60–3	Elevation 7500 feet west of upper White Creek near south boundary of medium-grained quartz monzonite; 49°53′N., 116°22′W.	Late body of medium-grained quartz monzonite	18

TABLE I (*Continued*)

Sample number	Reference number	Location	Rock type	Age
W-58-RA-19	60-4	Elevation 4300 feet in White Creek near boundary with hornblende–biotite granodiorite; 49°50'N., 116°17'W.	Porphyritic quartz monzonite	29
W-58-RA-22a	60-5	Elevation 4200 feet near southern boundary of batholith; 49°48'N., 116°16'W.	"Dioritized" inclusion in biotite grano-diorite	56
W-58-RA-9	60-6	Elevation 8000 feet east of White Creek in basin east of, and below, Skookumchuck Mountain; 49°51'N., 116°14'W.	Porphyritic quartz monzonite	60
W-58-RA-22	61-9	Elevation 4200 feet on White Creek near southern boundary; 49°48'33"N., 116°13'10"W.	Biotite granodiorite (encloses 22a, above)	73
W-58-RA-14	60-7	Elevation 7500 feet near southern boundary east of White Creek; 49°48'N., 116°13'W.	Hornblende–biotite granodiorite	79
W-58-RA-20	61-11 61-10	Elevation 5000 feet on east side of White Creek; 49°51'10"N., 116°16'40"W.	Leucoquartz monzonite core	80(M) 82
W-30-RA-1	62-2	1.88 miles south of White Creek batholith in top of very large Moyie sill; 49°47'18"N., 116°18'54"W.	Altered quartz diorite sill	126
W-32-RA-1	62-3	3.8 miles south of White Creek batholith in top of same sill west of White Creek; 49°45'42"N., 116°19'48"W.	Altered quartz diorite sill	138
		ANGUS CREEK STOCK		
LD-ML-5	62-43	Angus Creek; 49°33'N., 116°08'36"W.		118
		BAYONNE BATHOLITH		
Bay-5-RA-2	62-7	1.8 miles south of Kuskanook on Creston Highway; 49°16'30"N., 116°39'00"W.	Hornblende–biotite granodiorite	33
Baadsgaard *et al.*	KA 118	Kootenay Lake; 49°15'N., 116°39'W.	Granite or alkaline granodiorite	77
MW-52-RA-1	62-6	25.0 miles south of Kootenay Bay on Creston Highway; 49°24'30"N., 116°44'18"W.	Leucoquartz monzonite with megacrysts of potash feldspar in leucogranodiorite matrix	100
		PORCUPINE CREEK STOCK		
P-1-RA-1	62-5	0.35 miles from eastern contact; 49°15'N., 117°05'W.	Hornblende–biotite granodiorite with 5.8 per cent potash feldspar megacrysts	128

TABLE I (Continued)

Sample number	Reference number	Location	Rock type	Age
		LOST CREEK STOCK		
L-2-RA-1	62–4	4.9 miles up road from Salmo-Creston cut-off; 49°05′36″N., 117°10′12″W.	Coarse leucoquartz monzonite	119
		KUSKANAX BATHOLITH		
K-15-RA-2	62–33 62–34	2 miles south on road from Galena Bay; 50°41′N., 117°53′W.	Muscovite–biotite leucogranodiorite	66 90(M)
		NELSON BATHOLITH		
320-RA-1	60 21	0.3 mile south of junction of Duhamel Creek Road and Lemon Creek Road; 49°42′N., 117°19′W.	Quartz monzonite with 5 per cent biotite	49
321-RA-1	60 22	9.0 miles south of junction of Duhamel Creek Road and Lemon Creek Road; 49°36′N., 117°15′W.	Hornblende–biotite granodiorite	55
MW-35-RA-2	62–32	2.4 miles west from traffic signal in Nelson at west end of Baker Street; 49°29′18″N., 117°20′30″W.	Biotite granodiorite	63
Little(1960)	59–1	Quarry west of Nelson; 49°29′N., 117°20′W.	Granodiorite	86
MW-48-RA-1	62–31	10.9 miles east of Nelson Bridge; 49°36′36″N., 117°07′54″W.	Hornblende–biotite granodiorite	105
N-3-RA-1	61–17	Elevation 7500 feet south of Mt. Chipman; 49°51′30″N., 117°02′48″W.	Hornblende granodiorite	131
N-2-RA-1	62–27	Pontiac Peak; 49°46′24″N., 117°04′30″W.	Hornblende–biotite granodiorite	159
N-11-RA-1	62–29	10.2 miles from Slocan on Springer Creek Road; 49°46′54″N., 117°20′18″W.	Hornblende–biotite granodiorite	163
N-1-RA-1	62–26	Nansen Mountain; 49°45′48″N., 117°13′30″W.	Hornblende–biotite granodiorite	165
N-9-RA-1	62–28	Bridge on Springer Creek; 49°47′00″N., 117°22′24″W.	Hornblende–biotite granodiorite	171
N-16-RA-1	62–30	Elevation 7900 feet in basin north of Mt. Carlyle; 49°56′N., 117°08′W.	Hornblende–biotite granodiorite	171
		SHUSWAP METAMORPHIC COMPLEX		
		VALHALLA COMPLEX		
230-RA-2	59–5	South and above west end of Evans Lake; 49°51′N., 117°41′W.	Leucogranite gneiss (vein network of Veined Gneiss)	11

TABLE 1 (Continued)

Sample number	Reference number	Location	Rock type	Age
47-RA-4	59-4	1500 feet above east end of Evans Lake; 49°51'00"N., 117°37'30"W.	Biotite augen gneiss	13
582-RA-1	60-8	Head of Nemo Creek; 49°57'N., 117°35'10"W.	Biotite granite gneiss	15
51-RA-3	59-6	East end of Evans Lake; 49°51'30"N., 117°37'00"W.	Hornblende augen gneiss	16
602-RA-2	60-9	Head of Beatrice Valley; 49°52'30"N., 117°41'15"W.	Migmatite	25
353-RA-1	60 10	North of mouth of Russell Creek at first switchback in road; 49°37'N., 117°48'W.	Porphyritic leucoquartz monzonite gneiss	28
390-RA-1	60-11	Gwillim window; 49°47'30"N., 117°37'W.	Coarse biotite granite gneiss in Hybrid Gneiss	31
420-RA-3	60-12	East of headwaters of upper Koch Creek; 49°48'N., 117°49'40"W.	Leucogranite gneiss	42
506-RA-1	60-13	Mt. Rinda; 49°42'53"N., 117°42'20"W.	Pegmatite	46
F-13-RA-3	60-14	Edge of Veined Gneiss southwest of lower Gwillim Creek; 49°46'N., 117°29'W.	Biotite augen gneiss	47
608-RA-2	60-15	Head of Beatrice Valley; 49°53'13"N., 117°42'W.	Leucogranite gneiss	58
D-296-RA-1	61-16	South end of 8000 foot mountain top; 49°48'N., 117°54'W.	Porphyritic hornblende granite	58
D-386-RA-1	61-15	Elevation 4800 feet northeast of junction of Watson and Koch Creeks; 49°42'25"N., 117°54'W.	Biotite granite gneiss	59
535-RA-1	60-16	North of, and above, Beatrice Lake; 40°53'47"N., 117°37'20"W.	Medium- to coarse-grained biotite granite gneiss of Hybrid Gneiss	60
5-RA-4	60-17	North of mouth of Bannock Burn; 49°42'N., 117°36'W.	Hornblende augen gneiss	62
D-253-RA-1	61-14	Elevation 7000 feet southeast from northwest corner of Passmore map; 49°44'N., 117°58'W.	Biotite granite	66
HQ-6-1	62-39	Bridge across Cariboo Creek; 49°59'14"N., 117°50'00"W.	Porphyritic quartz monzonite	69
		MONASHEE GROUP		
Baadsgaard et al.	AK 25	Christina Lake; 49°01'N., 118°18'W.	Gneissic granite	36
637-RA-3	61-5	East shore of Mabel Lake, 5.4 miles south of Tsuius Creek; 50°34'N., 118°43'W.	Sillimanite–garnet–biotite–quartz–plagioclase gneiss	52

TABLE I (*Continued*)

le er	Reference number	Location	Rock type
	61–4	Eastern shore of Mabel Lake; 50°34′N., 118°43′W.	Pegmatite
d et al.	AK 29	10 miles north of Revelstoke; 51°08′N., 118°11′W.	Gneissic granodiorite
	60–1	2.4 miles west of Lavington Creek; 50°15′N., 119°09′W.	Calcareous gneiss—100 yards west of u conformity
	62–35	Lower end of glacier in cirque north of, and below, Mt. Odin; 50°33′12″N., 118°04′54″W.	Granodiorite gneiss
	62–45 62–44	Blanket Mountain; 50°49′13″N., 118°15′09″W.	Medium-grained biotite muscovite qua zite
	61–6	Three Valley Lake, 12 miles west of Revelstoke; 50°56′N., 118°27′W.	Pegmatite
	62–46	Blanket Mountain; 50°49′00″N., 118°15′30″W.	Coarse-grained tourmaline muscovite quartzite
	62–47	Blanket Moutain; 50°47′24″N., 118°15′16″W.	Coarse biotite granite
	61–8	North side of Blanket "syncline"; 50°49′03″N., 118°15′48″W.	Pegmatite cutting across all other stru tures
	62–48	Blanket Mountain; 50°46′44″N., 118°14′16″W.	Biotite–quartz–plagioclase paragneiss
1	62–36	Elevation 7500 feet on nose east of Mt. Odin; 50°32′N., 118°02′W.	Biotite–feldspar–quartz gneiss
	61–7	Three Valley Lake, 12 miles west of Revelstoke; 50°56′N., 118°27′W.	Biotite–quartz–feldspar paragneiss
d et al.	AK 27	Watshan Lake; 50°00′N., 118°06′W.	Granite
d et al.	AK 47	Monashee Pass; 49°58′N., 118°18′W.	Porphyritic granite

MT. IDA GROUP

-RA-1	62–37	Large road cut at west end of Salmon Arm of Shuswap Lake; 50°46′N., 119°20′W.	
	61–2 61–3	5.3 miles west of Squilax Bridge; 50°48′N., 118°42′W.	Biotite–muscovite schist
	61–1	2.8 miles west of Squilax Bridge; 50°47′30″N., 118°40′00″W.	Biotite–quartz–plagioclase gneiss

TABLE I (*Concluded*)

Sample number	Reference number	Location	Rock type	Age
GUICHON BATHOLITH				
Baadsgaard et al.	AK 44	Highland Valley; 50°29'N., 120°58'W.	Quartz diorite	186
CA-7-61	62-63	50°29'10"N., 121°03'20"W.	Granodiorite	224
CA-3-61	62-59	¼ mile northwest of Foot Lake; 50°21'25"N., 120°55'05"W.	Quartz diorite	227
CA-5-61	62-61	400 feet east of Outrider Road, 1 mile north of Highland Valley; 50°28'55"N., 120°55'40"W.	Quartz diorite (two separate argon extractions on same mineral concentrate)	230 237
CA-2-61	62-60	50°29'00"N., 120°55'00"W.	Quartz diorite	237
CA-6-61	62-62	50°30'00"N., 121°00'00"W.	Granodiorite	242
CA-4-61	62-58	50°30'03"N., 120°56'25"W.	Quartz diorite	245
CORYELL-TYPE GRANITIC ROCKS				
647-RA-1	60-20	20.1 miles west of Castlegar west of Deer Park; 49°24'N., 118°2'30"W.	Biotite–hornblende leucosyenite	27
D-341-RA-1	61-13	Elevation 7500 feet just below ridge top; 49°50'30"N., 117°56'30"W.	Hornblende–biotite granite	32
D-310-RA-1	61-12	8000 feet peak 1 mile northwest of Mt. Hilda; 49°49'15"N., 117°56'20"W.	Biotite granite	53
Baadsgaard et al.	AK 28	Santa Rose Pass; 49°03'N., 118°03'W.		54
Baadsgaard et al.	AK 26	Rossland, B.C.; 49°04'N., 117°58'W.		58

NOTES ON THE PLEISTOCENE TIME-SCALE IN CANADA

A. Dreimanis

ABSTRACT

The Pleistocene stratigraphy is based to a great extent upon climatic criteria because of repetition of extensive glaciations during this epoch. At present it is customary to adhere to the classical terrestrial stratigraphic divisions of the Pleistocene, but future stratigraphies will probably rely more upon continuous ocean-bottom records, dated by radiogenic isotope methods.

Most Canadian Pleistocene deposits are terrestrial, with lateral lithologic changes and numerous stratigraphic gaps. The most complete records have been found in the peripheral areas of the glaciations; several of them contain also marine deposits. Restrictions in methods of relative and absolute datings and correlations (rarity of fossils, uncertainties in weathering estimates and varve chronology, lack of absolute dates beyond the C-14 dating range, difficulties in long-distance lithologic correlations, etc.) have resulted in a more or less conjectural classification particularly of those Pleistocene deposits that are older than the last ice age. The post-Sangamon stratigraphy, aided by C-14 dates, has been developed with greater certainty, particularly in Ontario, southwestern Quebec, southwestern British Columbia, and the Canadian Arctic. The last ice age, including cool interstadials that lasted for thousands of years, was probably at least 70,000 years long.

THE LATEST and shortest period in the geologic time-scale is the Quaternary. Often it has been called the "last million years," for instance in the title of Coleman's last book (1941). Even though recent age determinations have placed the beginning of the Pleistocene at dates ranging from 600,000 years to close to 2 million years ago (see also p. 147), the Quaternary is considerably shorter than the preceding Tertiary period, which lasted for more than 60 million years (Kulp 1961, p. 1111). Therefore, most geologists quoted in this paper have preferred to drop the name Quaternary *period* and to use Lyell's (1839, quoted from Wilmarth, 1925, p. 47) term Pleistocene which has the rank of an *epoch*. Another common practice is to include the Recent in the Pleistocene, as it has been only approximately 10,000 years long or even shorter. This agrees also with Lyell's original meaning of Pleistocene as the entire post-Pliocene time (for details on the post-Pliocene terminology see Wilmarth, 1925, pp. 45–9, and Flint, 1957, pp. 278–81).

CLIMATOLOGIC STRATIGRAPHY

The main characteristic of the Pleistocene epoch is repetition of extensive glaciations. The glacial ages were separated by episodes, longer than the Recent, when the climate became similar to the present one, thus causing shrinkage and even disappearance of glaciers. Because of these alternating drastic climatic changes it has become customary to subdivide the Pleistocene into *ice ages* and *interglacial ages*. Both the old (Ashley *et al.* 1933) and the new Codes of Stratigraphic Nomenclature (American Commission on Stratigraphic Nomenclature, 1961) for this continent have recognized this peculiarity of the Pleistocene, that is, its *climatologic stratigraphy*, even though they differ in opinions as to whether glacial stages may be considered as time-stratigraphic terms. The old code, for instance, authorized the use of "Wisconsin glacial stage" as a time-stratigraphic unit. Although the new code (pp. 659–60) rejects the usage of such a term as a time-stratigraphic subdivision, it still recognizes "Wisconsin glaciation" as a geologic climate unit, but notes that it is time-transgressive.

The Pliocene–Pleistocene boundary is based also upon climatic criteria; for instance, the Eighteenth International Geological Congress (1950) recommended that the Pliocene–Pleistocene boundary be defined by the first appearance of the cool-climate foraminifera *Anomalina baltica* in the Mediterranean region. Ericson *et al.* (1963) have found recently, by studying deep-sea sediments, that a relatively abrupt and world-wide cooling of climate caused pronounced extinctions and evolutionary changes among planktonic fossils at the transition from Pliocene to Pleistocene. As these changes occurred during a time interval of less than 6000 years (estimated by Ericson *et al.* 1963), a distinct biostratigraphic boundary was developed in ocean sediments.

Biostratigraphic units, similar to those of the pre-Pleistocene sedimentary record, have been used as the basis for geologic time subdivisions of the Pleistocene, particularly in non-glaciated areas (see Flint, 1957, pp. 446–72, for a review). However, it is difficult to apply the biostratigraphic units to sections where fossils are absent or rare, for example, to most of the glaciated terrain.

Many of the Pleistocene evolutionary changes have not been simultaneous on all continents. Thus a great number of biostratigraphic boundaries tend to be time-transgressive. The time difference, even if measurable in a few thousands or tens of thousands of years, may be significant, if we consider the short span of the Pleistocene epoch.

As many of the recognizable evolutionary changes were due to climatic causes, thus, even from the biostratigraphic viewpoint, climate has to be considered as the most important factor in Pleistocene stratigraphy and correlations. It is hoped that a combination of biostratigraphic, rock-stratigraphic, and climatic criteria, supplemented by isotopic dates, will form a basis for a well-founded Pleistocene stratigraphy in the future.

Man

The appearance and development of man during the Quaternary was such a significant event that some scientists prefer to replace the term

Quaternary by a new one more related to man: Anthropogen (see Gromov *et al.* 1960, p. 13). On our continent man arrived only towards the end of the Pleistocene: immigration of humans from Asia to North America, via the Bering Strait, began most probably not earlier than during the last ice age, though some archaeologists consider also the penultimate ice age as the beginning of this immigration. The southeastward spread of man across our continent occurred probably during the Main or Mid-Wisconsin interstadials and Postglacial time (see MacNeish, 1951 and 1956, Wormington, 1957; and Griffin, 1960, for summaries on archaeological findings and further references on Canada and the adjoining parts of the United States).

The oldest radiocarbon-dated human sites (more than 20,000 years B.P.) are known from the western United States. If man arrived there via the Bering Strait, he must have traversed Western Canada. Some of the most likely ancient migration routes that deserve particular attention of students are along the Mackenzie and the Yukon valleys and the western plains and foothills east of the Rocky Mountains.

The Great Lakes region was reached by Palaeo-Indians approximately 11,000 to 10,500 year ago (Mason 1960, p. 374).* Dating of the early artefacts and sites has been facilitated particularly in areas where they were associated with datable ancient beaches and other lacustrine or marine deposits (see Greenman and Stanley 1943; MacNeish 1952; Elson 1957*a*; Wright 1963; etc.). Therefore, archaeologists are much concerned with the history of the Late- and Post-Glacial lake and marine stages (see, for instance, Mason 1960). In the Old World even some complete Pleistocene stratigraphies have been based upon a combination of geological and archaeological information.

Spreading of man throughout this continent coincides with the disappearance of several species of large mammals, such as mastodon, mammoth, camel, horse, and some bisons, during the Late Wisconsin and Early Recent times. This extinction came to a climax about 8000 years ago (Hester 1960). Extensive hunting and epidemic diseases may have been some of the main causes; further investigations are still required.

Most of the archaeological findings in Canada are related to Recent time and therefore will not be discussed in detail (see also p. 153).

CLASSICAL MIDWEST STRATIGRAPHY

For a long time the *terrestrial* Pleistocene record of the upper Mississippi valley area has been considered as the most complete on this continent. As a result, the time-stratigraphic terms that were established by T. C. Chamberlin, F. Leverett, and others in this so-called Midwest area (see Table I) became a standard for the central, northern, and eastern portions of North America, including Canada.

*Lee (1957) suggests the appearance of man on Manitoulin Island more than 30,000 years ago, but much controversy exists about this very early date.

TABLE I

THE PLEISTOCENE STRATIGRAPHY (STAGES)
OF THE MIDWEST AREA OF NORTH AMERICA
(AFTER FLINT 1957, p. 336)

Wisconsin glacial
Sangamon interglacial
Illinoian glacial
Yarmouth interglacial
Kansan glacial
Aftonian interglacial
Nebraskan glacial

Although the total number of glacial and interglacial ages, plus the Recent, may have been greater than eight, these eight major subdivisions have been generally accepted. They also agree with a similar number of glacial and interglacial ages in the classical areas of Europe.

PRESERVATION OF DEPOSITS OLDER THAN LAST ICE AGE

To the best of the author's knowledge, there is no single geological section in the classical Midwest or any other glaciated area where a complete record of the eight Pleistocene ages has been preserved. Each glacial advance wiped out most of the sediments of the previous glacial and interglacial ages. Other terrestrial erosional agents contributed to this destruction of the Pleistocene records. As a result, in most glaciated areas, particularly in Canada, the bedrock is covered by deposits of the last ice age and the Recent only. Exceptions are in the peripheral zones of the last glaciation, where the probability is greatest that non-consolidated materials, older than the last ice age, have not been eroded.

The following Canadian areas that were peripheral during the last glaciation or were not glaciated may be considered most promising for the findings of Pleistocene deposits older than the last ice age (see Fig. 1):

(a) the nonglaciated area in the Yukon and Northwest Arctic, and the Arctic peripheral glaciated zone;

(b) the dividing zone between the Cordilleran and the Labradorean ice sheets, where several foothill areas have not been glaciated;

(c) southern Alberta and Saskatchewan, where some of the high hills have also been ice-free.

Inside the ice sheet cover, some of the older Pleistocene deposits may have been preserved in locally protected areas:

(a) in buried old valleys, e.g. in southern Alberta (Stalker 1960, p. 70; 1961, pp. 4–5),

(b) in sheltered places behind bedrock ridges or steps,

(c) along the confluence of two opposing glacial flows, if they slowed down each other's velocity and erosive power,

(d) in cores of thrust moraines, e.g. the Port Talbot interstadial type section (Dreimanis 1958).

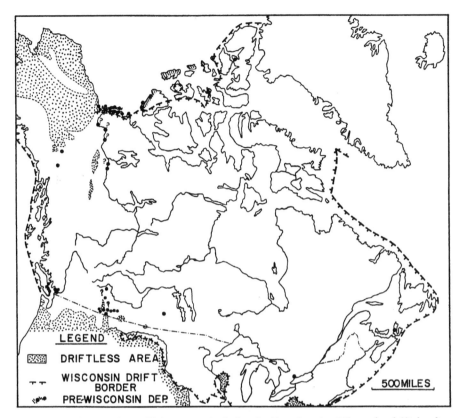

FIGURE 1. Extent of the Pleistocene glaciation in Canada, the Wisconsin drift border, and most important occurrences of pre-Wisconsin deposits, listed on pp. 148–9. Boundaries from Prest (1961, map 1) and Glacial Map of Canada (1958).

Summary accounts and brief descriptions of multiple drifts, deep weathering, buried peat deposits, and other sediments or volcanics in Canada, probably older than the last ice age, have been published in several recent reports, which contain also lists of older references. The following may be noted: Prest (1957, pp. 446–61; 1961, pp. 8–13), Armstrong (1961, pp. 23–5), Christiansen (1960, p. 52), Craig and Fyles (1960, pp. 2–5), Stalker (1963, p. 3), Terasmae (1958, 1960). Most of the occurrences mentioned in the above papers are shown in Figure 1, but their descriptions will not be repeated here.

LITHOLOGIC CORRELATIONS

The surface deposits can be traced continuously. It is more difficult to interpret, date, and correlate the widely scattered older Pleistocene beds because of the extensive gaps between them, the absence of fossils in most sections, and the lack of detailed knowledge of regional changes in the lithology of each drift sheet. Differences in lithology and texture have made it possible to distinguish tills according to their provenance, for example to differentiate various lobal areas in the Great Lakes region (Dreimanis 1961), to distinguish the Cordilleran from the Laurentide drift in western Alberta (Horberg 1954; Rutulis 1962), or the Patrician from the Keewatin drift in Manitoba (Elson 1961). On several other occasions, it has been difficult to distinguish tills of different provenance and different age.

Fossil Remains

Palynologic studies have been found most useful, particularly for ecologic and climatologic interpretations of non-glacial sediments (see Terasmae 1958, 1960, 1961). Because of the scarcity of interglacial deposits, more investigations should be done on the pollen content in the basal portions of till layers, similar to Heinonen's (1957) studies in Europe. These may help in reconstructing the records of the eroded interglacial and interstadial deposits.

Fossil remains of invertebrates and vertebrates have been studied or listed from various last ice age deposits, for example by Elson (1960), Russell (1923, 1956), Sternberg (1930, 1963), and Wagner (1959). Less is known about the Canadian pre-Wisconsin animal fossils, even though their remains have been reported to be quite abundant in the non-glaciated area in the Yukon.

Weathering Profiles and Erosional Surfaces

Examination of weathering profiles (soils) and erosional surfaces in the periphery of glacial cover has suggested that some of these areas were subjected to weathering and erosion for a considerably longer time than others. Stalker (1959) concluded that the oldest-looking glacial surface deposits in the Fort Macleod map area, southwestern Alberta, were even of Nebraskan or Kansan age. Further detailed studies of these weathering surfaces, in co-operation with pedologists, have been suggested for cross-checking this proposed very old Pleistocene age.

Weathering profiles, buried underneath one or several tills or other sediments, have been found in a few scattered locations in Canada, for example at Toronto (Watt 1954; reinterpreted by Dreimanis and Terasmae 1958) and in Alberta, (Horberg 1954; Stalker 1963), but none of these appear to be as much weathered as the interglacial palaeosols south of the international boundary. This may be due to differences in climate; or the Canadian palaeosols known at present may be interstadial rather than interglacial. The extensive intertill boulder pavements in Manitoba and Saskatchewan (Elson 1957b) probably represent fluctuations of the glacial margin during the last ice age.

Depth of leaching of soils has also been used for estimating the absolute length of time required for weathering. When the names of the Pleistocene glacial and interglacial ages were proposed, no absolute dates were assigned to them. Attempts at absolute dating by various methods followed soon. Among the time estimates, the most-quoted figures have been those of Kay (1931), before the isotopic dates became available. He inferred the length of interglacials from the depth of carbonate leaching in palaeosols, and the length of ice ages from average figures of glacial advances and retreats from Iowa. His interglacials have been estimated as too long, as he assumed 25,000 years for the formation of the Recent soils in Iowa (the radiocarbon dates suggest half of this time).

Estimation of the length of weathering in absolute years has been questioned repeatedly. However, with some caution, and by comparing soils that have been formed under identical circumstances, it may be possible to use soils as criteria of absolute time and for telecorrelations of beginning of weathering (meaning retreat of glaciers or lowering of proglacial lakes), even across the Atlantic Ocean (Dreimanis 1959). Geologists still need the assistance of pedologists and soil chemists in order to draw correct stratigraphic conclusions from soil profiles.

VARVE CHRONOLOGY

Varves, if correctly identified, represent annual* periods of deposition. By correlating them from one proglacial deposit to another on short distances and along the line of general glacial retreat, De Geer, Lidén, Sauramo, and other Swedish and Finnish geologists have succeeded in developing absolute chronologies for the Late- and Post-Glacial time in the Baltic region (see Antevs, 1953, and Järnefors, 1963, for brief summaries and further references). The Finno-Swedish varve chronology, except for its earliest portion, agrees reasonably well with radiocarbon datings.

Antevs (see 1953 and 1957, and his other papers listed therein) has developed a varve chronology for eastern North America. Considerable discrepancy exists between his chronology and radiocarbon dates. For instance, he proposes that the Valders maximum was 19,000 years B.P., as compared with approximately 11,000 years, derived from C-14 dates. The principal difference between the Finno-Swedish and Antevs' North American varve chronologies is that the former have developed from almost continuous varve counts up to historical times, while "the varve chronology of North America is based on discontinuous varve measurements . . . and on geologic estimates of the gaps in the varve data" (Antevs 1957, p. 129). No local varve record connects Antevs' chronology with historical times, and he has assumed that the ice oscillations at Cochrane (the latest events in Antevs' chronology) are correlative with the Nyland and Salpausselkä halts in Finland. The restudy by Terasmae and Hughes (1960) of some of Antevs' varve measurements in Ontario confirms that Antevs' counting

*Distinction of non-seasonal laminae from varves will not be discussed in this paper.

and interpretation of varves as annual deposits were correct. However, considerable uncertainty remains about the estimates of length of the un-measured gaps in Antevs' chronology. Several of them appear to be too long, particularly those which correspond to rapid glacial retreats due to calving in the extensive proglacial lakes. The Cochrane–Salpausselkä correlation may also be questioned. Transoceanic geologic correlations, including the much-disputed G. De Geer's and E. H. De Geer's telecorrelations (see E. H. De Geer, 1962, and the papers referred to therein) always include more assumptions and uncertainties than local correlations.

In view of the uncertainties in time estimates of the unmeasured gaps in the North American varve chronology and its correlation with the Finno-Scandinavian varve record, it may be safer to use only those portions of the North American varve chronology that have been based upon actual varve measurements and short-distance correlations. This means that the North American varve chronology is still fragmentary, floating in time, and has to be tied to other absolute time-scales.

ISOTOPIC AGE DETERMINATIONS

After isotopic age determination methods became available in dating Pleistocene events, absolute dates could be assigned to some of the relative stratigraphic units. They have served several purposes, for example:

(a) determination of the length of established time-stratigraphic units,
(b) finding of new, previously unknown, stratigraphic units,
(c) correlation of rock-stratigraphic units by their absolute dates.

The most used and most popular is the C-14 dating method (Libby 1955) of organic remains, even though a few scientists (Antevs 1953, 1957) are sceptical about it. Unfortunately it can be applied only as far back as 70,000 years with the methods now available, and in most laboratories only as far as 40,000. This dating range does not reach as far as the last interglacial age, but the precision of the C-14 dating has made this method suitable as an important criterion in studies of the last ice age. Radiocarbon dates are used even in descriptions of time-stratigraphic units, because the absolute C-14 date is often the most distinctive feature of two otherwise similar units. This reliance upon C-14 dates in Pleistocene stratigraphy is much greater than the reliance upon isotopic dates applied to older geologic events.

In so far as terrestrial Pleistocene deposits are concerned, none of the isotopes used for age determinations older than the radiocarbon dating range have given the same degree of accuracy as C-14.

The potassium–argon method has been found suitable for dating volcanic sanidines of Pleistocene age (Evernden 1959). The half-life of K^{40} is 1.3×10^9 years, and considerable error may be caused by contamination or loss of argon. Most of the reported Pleistocene potassium–argon dates tend to be greater than those obtained by other isotopic methods (though from different samples). Thus, the latest K/Ar dates from volcanics, associated with the early human remains in non-glaciated Africa, place the beginning of the Pleistocene at more than 1.75 million years ago (Leakey *et al.* 1961, p. 479); and Everden (1959, p. 19) had obtained a K/Ar date of a million years for the Bishop tuff, intercalated between tills in Sierra Nevada, California.

With further improvements of the potassium–argon method absolute dates of the Early and Mid-Pleistocene volcanics, including those found in Western Canada, may become significant time-markers in the Pleistocene chronology.

Good progress has been made in chronologic studies of ocean-bottom sediments. Several isotopes—ionium (Th^{230}), radium (R^{226}), protactinium (Pa^{231}), and others—have been used during the last 25 years for absolute dating of ocean cores (see Rosholt *et al.*, 1961, for a brief review and references). Rosholt *et al.* (1961) claim that the ratio of protactinium (Pa^{231}) to ionium (Th^{230}) is most reliable in dating undisturbed ocean-bottom sediments. These two radionuclides are produced by decay of uranium, but their own decay is independent and with different half-lives. The "half-life" of the ratio is approximately 60,000 years, which makes it very suitable for dating ocean-bottom Pleistocene deposits beyond the C-14 range.

Most absolute dates obtained by the protactinium–ionium method are up to 50 per cent smaller than dates based upon other isotopic methods or assumed rates of sedimentation (compare the dates of Rosholt *et al.*, 1961, or Emiliani, 1961, with those of Ericson *et al.*, 1963). Further studies, tests, comparisons, with other criteria are still required.

The Deep Sea Stratigraphic Record

The ocean-bottom sediments show biogeologic, geochemical, and sedimentologic evidences of climatic fluctuations similar to those obtained from terrestrial records: alternating colder and warmer episodes. There is a greater chance of obtaining continuous sedimentary columns for the entire Pleistocene in the ocean bottom than on land, and radiogenic isotope methods, as mentioned before, are available for assigning absolute ages to these records (see, for example, Emiliani, 1961, p. 530). Thus a prediction may be justified that a detailed Pleistocene time-scale will be worked out from ocean-bottom records, and all the terrestrial stratigraphies will have to be correlated with marine standard, particularly in that portion of the Pleistocene which is beyond the radiocarbon dating range.

Areas for which the layers of terrestrial and marine sediments overlap one another, for instance the St. Lawrence Lowlands, the Arctic, and the west coast of British Columbia, will provide an opportunity to check the correlation of at least some of the terrestrial and marine records.

Regional comparison of the isotope-dated ocean-bottom records will help also to decipher the causes of the Pleistocene climatic fluctuations. For instance, a comparison of dated temperature changes in the Arctic and Atlantic Oceans may tell whether Ewing and Donn (1956) were right in postulating that the Arctic Ocean was successively warmer and ice-free at the beginning of each ice age.

A brief review on the present status of the Pleistocene time-scale in Canada will follow this general discussion of some of the significant problems and concepts in the Pleistocene stratigraphy.

Deposits Older than the Last Ice Age in Canada

Because of lack of absolute dates and incomplete fossil records it is very difficult to assign definite time-stratigraphic names to those Canadian Pleistocene deposits that are older than the last or Wisconsin ice age. All the stratigraphic correlations have been tentative, and the age assignments have varied from one author to another. Thus, according to Stalker (1960, p. 60; 1963, p. 3) drifts of all the major four ice ages may be present in Alberta. He mentions, however, that "there is complete disagreement in the conclusions of the various observers" (see Stalker, 1960, p. 68, and the list of references, p. 67), for example, "estimates of the age of the lowest (Labuma) drift alone range from Nebraskan to Wisconsin."

The absence of palaeosols between tills, and the rareness of organic deposits that can be regarded without any doubt as interglacial rather than interstadial, are main reasons why considerable differences of opinions exist about the possible age of the multiple drifts in southern Alberta.

Armstrong (1961, p. 24), while summarizing the sequence of the Pleistocene deposits in the coastal area of southwestern British Columbia, assigns a local name, Pre-Seymour Group, to the lowest glacial, intertill, and preglacial sediments, which are up to 1500 feet thick and appear to be older than the last ice age. He states, however, that insufficient evidence is available to determine the actual sequence.

Those northwestern areas of the Arctic islands and the Yukon which were not glaciated during the last ice age contain wide-spread deposits, older than the last ice age (Prest 1961, pp. 11–12; Craig and Fyles 1960, pp. 2–5). Because of the relative abundance of sections and the variety of fossils, this region appears to be very promising for development of a pre-Wisconsin stratigraphic column. The Early Pleistocene record may be deciphered from the Beaufort formation that is late Tertiary and early Pleistocene in age (Tozer 1956, pp. 25–28; Craig and Fyles 1960, pp. 2–4). Excellent sections of Pleistocene deposits, younger than the Beaufort formation, have been found on the western Banks Island (Prest 1957, p. 457; 1961, p. 11; Craig and Fyles 1960, p.4). Though Prest (*ibid.*) considers the interglacial pond silts containing peaty material at Cape Kellett (Banks Island) to be of Aftonian or Yarmouth age, a younger, Sangamon, age is not excluded.

In order to avoid much speculation and subjective interpretation, no further attempt will be made in this paper to classify all the probable pre-Sangamon deposits in Canada.

Already less controversial information is available about some deposits of the last interglacial or Sangamon age, even though absolute dates are lacking.

The well-known Toronto Don beds are now generally considered to be of Sangamon age (see Terasmae, 1960, pp. 23–40, for the latest studies and a list of earlier references), but the overlying Scarborough beds are probably younger (*ibid.*, p. 39; see also Fig. 2).

In eastern Canada Sangamon age has been assigned to a buried peat and gyttja layer, and wood in till on Cape Breton Island (Prest 1961, p.9).

In Saskatchewan the Echo Lake Gravel, found betwen two tills along the south bluff of the Qu'Appelle valley between B-Say Tah and Fort Qu'Appelle (Christiansen 1960, pp. 33–34), may belong to the Sangamon interglacial according to L. S. Russell's conclusions from the vertebrate fauna.

Several of the pre-Wisconsin organic deposits of western and northwestern Canada mentioned on page 148 may be also of Sangamon age. After more palynologic investigations and studies of other fossil remains become available, all the regionally scattered Sangamon deposits will give a much better idea about climate, flora, and fauna of the last interglacial age than we have at present. Studies of the Don bed flora have repeatedly led to a climatological conclusion that "the annual mean temperature at the time of their deposition reached a maximum probably 5° F warmer than the present" (Terasmae 1960, p. 37).

WISCONSIN OR LAST ICE AGE

Palynologic investigations of several new interstadial beds and lithologic studies of tills in southern Ontario and southwestern Quebec, supplemented by radiocarbon dates as far back as 66,000 years B.P. (Dreimanis 1960a, 1960b, and 1961; Gadd 1960; Terasmae 1958, 1960, and the references therein), have made it possible to develop a new stratigraphic column for the last ice age in the eastern Great Lakes – St. Lawrence River region. According to Flint (1963, pp. 402 and 403) this area is now providing the most detailed record on the pre-classical Wisconsin post-Sangamon sediments, and can serve as a standard for comparison. Frye and Willman (1963, pp. 14 and 17) are also correlating their loess stratigraphy of the classical Midwest area with the Ontario sequence. Leighton (1963, p. 1) even suggests Ontario as an appropriate name for the early glacial substage of the Wisconsin Glacial Stage, because the ice sheet advanced over Ontario towards the southwest (for lithologic proof of this movement, see Dreimanis, 1960b, Fig. 1).

Figure 2 shows a chart of the possible major glacial advances and retreats

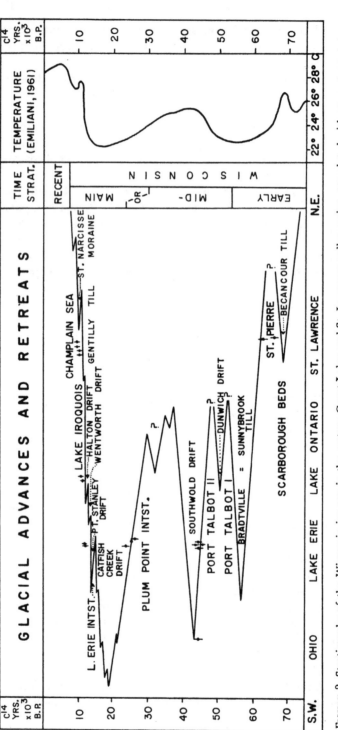

Figure 2. Stratigraphy of the Wisconsin ice age in the eastern Great Lakes and St. Lawrence valley region, correlated with ocean temperatures (Emiliani 1961). The glacial advances and retreats are projected upon a northeast southwest line from southern Labrador to central Ohio (approximately 1300 miles). Names of glacial deposits are given on the right side of the zigzag line of glacial fluctuations, and non-glacial terms on the left. Radiocarbon dates are plotted as dots, with their error range as vertical lines. The chart is based upon Dreimanis (1958, 1960a, 1960b, 1961), Flint (1963), Elson (1960), Goldthwait (1958), Gadd (1960), Karrow (1962), Terasmae (1958), 1960), and the author's unpublished studies, and, since it is only a progress report, it is subject to revision. *Note:* Only one of the C-14 dates, shown along the Southwold drift / Port Talbot II boundary, is from wood found in the Southwold drift; the other three are from the Port Talbot interstadial deposits.

along a line from southern Labrador southwestward over the Lake Ontario and the Lake Erie basins to north-central Ohio. Several of the glacial advances and non-glacial intervals have been dated by the radiocarbon method, some of the drifts have been traced by lithologic correlations, but large portions of the chart, particularly the end points of the glacial retreats, are merely tentative assumptions. The author hopes that some of the uncertainties will be clarified by studies in progress in the Port Talbot area and west of Toronto.

The lower portion of Figure 2 differs from previously published similar charts (Dreimanis 1960a, Fig. 3; Flint 1963, Fig. 2). According to the new findings in the two 60 to 130 foot deep test holes at the Port Talbot interstadial site on the north shore of Lake Erie,* the Port Talbot interstadial (now renamed tentatively as Port Talbot II) was preceded by a glacial advance of the Georgian Bay lobe reaching only as far as the northern shore of Lake Erie. This limited advance (not as extensive as previously assumed—see Dreimanis 1960a, p. 116) deposited the dolomite-rich Dunwich drift. The Dunwich drift, consisting mostly of glacio-lacustrine and ice-rafted deposits, is underlain by interstadial deposits of the Port Talbot I. This interstadial was preceded by a major glacial advance from east-northeast which deposited the reddish Bradtville drift (new name); this drift is probably correlative with most of those Early Wisconsin drifts that were correlated with the Dunwich drift previously (Dreimanis 1960a, pp. 113–6). The post-Port Talbot glacial advance which deposited the Southwold drift could be more extensive than previously assumed, if the Gahanna, Ohio, wood, 46,000±2,000 years old (Goldthwait 1958, p. 211; Flint 1963, p. 403), is related to this till (see Fig. 2). If this tentative new correlation is not correct, then merging of the Port Talbot II with the Plum Point intervals in one interstadial should be also considered.

More details on the pollen and other fossil content of the St. Pierre, the Port Talbot II, the Scarborough, and the Missinaibi peat beds (probably correlative with the St. Pierre), can be found in Dreimanis (1958, pp. 73–7) and Terasmae (1958, 1960).

When the glacial fluctuations of the eastern Great Lakes region are compared with Emiliani's (1961) average temperature record from deep sea sediments (see Fig. 2), several similarities and also discrepancies are noticeable; there are more glacial advances and retreats on land than major temperature changes in the ocean, and no lowering of ocean temperature corresponds to that glacial advance which deposited the Southwold drift. However, it is possible that the glacial advances were delayed, because of the necessary build-up period of ice, in comparison with the more rapid temperature changes in the oceans.

Detailed records of multiple glacial advances and retreats during the Wisconsin ice age have been reported also from southwestern British Colum-

*Research project supported by a grant from the Geological Survey of Canada.

TABLE II

CORRELATION CHART OF SOME TIME-STRATIGRAPHIC AND ROCK-STRATIGRAPHIC CLASSICATIONS OF THE WISCONSIN ICE A(
(The Recent and part of the Sangamon interglacial are included in the chart. The intraglacial names are left out of the Leight(
(1963) classification between Valders and Iowan because of lack of space.)

C¹⁴ years B.P.	N. America (Prest 1957, 1961)	E. Canada (Dreimanis 1960a, revised)	S.W. Brit.-Col. (Armstrong 1961)	Vancouver (Fyles 1963)	Midwestern U.S. (Frye and Willman 1960)	Midwestern U.S. (Leighton 1963)	De for N (E
	Recent	Recent	Salish and Capilano Groups	Salish and Capilano sediments	Recent	Recent	"R
10,000					Valderan Twocreekan	Valders	Wi
		Main Wisconsin	Sumas Group			Mankato	
				Vashon drift	Woodfordian	Cary	
20,000			Vashon Group			Tazewell	
	Wisconsin II		Semiamu Group		Farmdalian	Iowan	Wi
30,000		or				Farm Creek	
			Quadra Group	Quadra sediments		Farmdale	
40,000		Mid- Wisconsin				Pro- Farmdale time	
50,000	Mid-Wisconsin "break"				Altonian	–	T
60,000		Early Wisconsin	Seymour Group	Dashwood drift		eastern glacial lobes	F
70,000							Wi
80,000	Wisconsin I						
90,000					Sangamonian	Sangamon interglacial	San
		Sangamon	Pre-Seymour Group	Mapleguard sediments			

bia (Armstrong, 1961, pp. 23–24, and references therein; Fyles, 1963, pp. 14–100). These deposits of two major glacial advances of continental ice-sheet proportions, the Vashon and the Dashwood or Seymour drifts, are separated by extensive sediments of the cool-temperate climate Quadra Group. Though Armstrong (1961, p. 24) considers the Quadra Group to be probably interglacial, Fyles (1963a, pp. 37–8), Prest (1961, p. 15), and Wagner (1959, p. 52) are more in favour of a cool-temperate inter-stadial. As the radiocarbon ages of the Quadra sediments range from 25,000 to more than 41,500 years B.P., it is possible that they correspond to the entire Mid-Wisconsin time in the Great Lakes region, including both the Plum Point and the Port Talbot interstadials of southern Ontario. Table II suggests a tentative correlation between the areas.

The abundance of exposures containing the Quadra and other younger and older Wisconsin sediments in southwestern British Columbia is prom-ising for further development of a standard stratigraphic section for the west coast, providing that more finite radiocarbon dates become available.

Thus at least two standard stratigraphies for the last ice age are develop-ing in Canada; one in southwestern British Columbia, another in the eastern Great Lakes – St. Lawrence region. For regional correlations a minimum of one equally detailed section will be required for the Maritimes, two for the area between the Rocky Mountains and Lake Superior (southwestern Alberta with its multiple drifts is most promising), and several for the Canadian Arctic (at present the best chances are in the northwestern Arctic), for instance on Victoria Island, whence Fyles (1963b) reports intertill tundra vegetation, 28,000 radiocarbon years old.

The Postglacial and even the Lateglacial time will not be discussed in this paper even though some of the Lateglacial events are shown on Figure 2. A discussion of this time interval in Canada, with emphasis on climatic changes, has been published by Terasmae recently (1961).

SUMMARY

Most Canadian Pleistocene deposits are terrestrial, with lateral lithologic changes and numerous stratigraphic gaps. Correlation and classification of the pre-Sangamon and the Sangamon beds are more or less conjectural, mainly because of lack of absolute age determinations and rarity of fossils suitable for faunal classification. Palynological investigations and evalu-ation of the degree of weathering have been used with variable success. Most detailed information is available about the Toronto Don beds.

Several C-14 dated new subdivisions of the post-Sangamon stratigraphy have been established in Ontario, southwestern Quebec, and southwestern British Columbia (Table II and Fig. 2). A revised post-Sangamon time-scale could be based upon these and additional data and correlations with isotope-dated ocean-bottom cores.

154 A. DREIMANIS

REFERENCES

AMERICAN COMMISSION ON STRATIGRAPHIC NOMENCLATURE (1961). Code of stratigraphic nomenclature. Bull. Am. Assoc. Petrol. Geologists, 45: 645–65.
ANTEVS, E. (1953). Geochronology of the deglacial and neothermal ages. Jr. Geol., 61: 195–230.
——— (1957). Geological tests of the varve and radiocarbon chronologies. J. Geol., 65: 129–48.
ARMSTRONG, J. E. (1961). Soils of the coastal area of southwest British Columbia. In Soils in Canada, edited by R. F. Legget (Roy. Soc. Can., Spec. Publ. no. 3). pp. 22–32. Toronto: University of Toronto Press.
ASHLEY, G. H. and others (1933). Classification and nomenclature of rock units. Bull. Am. Assoc. Petrol. Geologists, 14: 423–59.
CHRISTIANSEN, E. E. (1960). Geology and ground-water resources of the Qu'Appelle area, Saskatchewan. Sask. Res. Council, Geol. Div. Rept. no. 1.
COLEMAN, A. P. (1941). The Last Million Years. Toronto: University of Toronto Press.
CRAIG, B. G. and FYLES, J. G. (1960). Pleistocene geology of arctic Canada. Geol. Surv. Can., Paper 60-10.
DEGEER, E. H. (1962). G. DeGeer's chronology. Geokronol. Inst., Univ. Holmiensis, Data 119.
DREIMANIS, A. (1958). Wisconsin stratigraphy at Port Talbot on the north shore of Lake Erie, Ontario. Ohio Jr. Sci., 58: 65–84.
——— (1959). Measurements of depth of carbonate leaching in service of Pleistocene stratigraphy. Geol. Fören. Förhandl., 81: 478–82.
——— (1960a). Pre-classical Wisconsin in the eastern portion of the Great Lakes region, North America. Intern. Geol. Congr. XXXI Sess., Norden, 1960, IV: 108–19.
——— (1960b). Supplement to "Preclassical Wisconsin in the eastern portion of the Great Lakes region, North America." Dept. Geol., Univ. Western Ont., Contrib. 33a.
——— (1961). Tills of southern Ontario. In Soils in Canada, edited by R. F. Legget (Roy. Soc. Can., Spec. Publ. no. 3), pp. 80–96. Toronto: University of Toronto Press.
DREIMANIS, A. and TERASMAE, J. (1958). Stratigraphy of Wisconsin glacial deposits of Toronto area, Ontario. Proc. Geol. Assoc. Can., 10: 119–35.
ELSON, J. A. (1957a). Lake Agassiz and the Mankato-Valders problem. Science, 126: 999–1002.
——— (1957b). Striated boulder pavements of southern Manitoba, Canada. Bull. Geol. Soc. Am., 68: 1722.
——— (1960). Littoral mollusks of the Champlain sea. Mimeographed.
——— (1961). Soils of the Lake Agassiz region. In Soils in Canada, edited by R. F. Legget (Roy. Soc. Can., Spec. Publ. no. 3), pp. 51–79. Toronto: University of Toronto Press.
EMILIANI, C. (1961). Cenozoic climatic changes as indicated by the stratigraphy and chronology of deep-sea cores of globigerina-ooze facies. Ann. N.Y. Acad. Sci., 95: 521–36.
ERICSON, D. B., EWING, M. and WOLLIN, G. (1963). Pliocene–Pleistocene boundary in deep-sea sediments. Science, 139: 727–37.
EVERNDEN, J. F. (1959). First results of research on the dating of Tertiary and Pleistocene rocks by the potassium/argon method. Geol. Soc. London Proc., Sess. 1958–59, 1565: 17–19.
EWING, M. and DONN, W. L. (1956). A theory of the ice ages. Science, 123: 1061.
FLINT, R. F. (1957). Glacial and Pleistocene Geology. New York: J. Wiley and Sons.
——— (1963). Status of the Pleistocene Wisconsin stage in central North America. Science, 139: 402–4.
FLINT, R. F. and BRANDTNER, F. (1961). Outline of climatic fluctuations since the last interglacial age. Ann. N.Y. Acad. Sci., 95: 457–60.

FRYE, J. C. and WILLMAN, H. B. (1960). Classification of the Wisconsin stage in the Lake Michigan glacial lobe. Ill. State Geol. Surv., Circ. 285.
——— (1963). Loess stratigraphy, Wisconsinan classification and accretion-gleys in central western Illinois. Ill. State Geol. Surv., Guidebook Ser. 5.
FYLES, J. G. (1963a). Surficial geology of Horne Lake and Parksville map-areas, Vancouver Island, British Columbia. Geol. Surv. Can., Mem. 318.
——— (1963b). Surficial geology of Victoria and Stefansson islands, District of Franklin. Geol. Surv. Can., Bull. 101.
GADD, N. R. (1960). Surficial geology of the Bécancour map-area, Quebec. Geol. Surv. Can., Paper 59-8.
GLACIAL MAP OF CANADA (1958). Published by Geol. Soc. Can.
GOLDTHWAIT, R. P. (1958). Wisconsin age forests in western Ohio. I. Age and glacial events. Ohio J. Sci., 58: 209–30.
GREENMAN, E. F. and STANLEY, G. M. (1943). The archaeology and geology of two early sites near Killarney, Ontario. Pap. Mich. Acad. Sci., 28: 505–31.
GRIFFIN, J. B. (1960). Some prehistoric connections between Siberia and America. Science, 131: 801–12.
GROMOV, V. I., KRASNOV, I. I., NIKIFOROVA, K. V., and SCHANZER, E. V. (1960). Principles of a stratigraphic subdivision of the Quaternary (Antropogen) system and its lower boundary. Intern. Geol. Congr. XXI Sess., Norden, IV: 7–26.
HEINONEN, L. (1957). Studies on the microfossils in the tills of the North European glaciation. Acad. Sci. Fenn. Ann., Ser. A, III, 52: 1–92.
HESTER, J. J. (1960). Late Pleistocene extinction and radiocarbon dating. Am. Antiquity, 26: 58–77.
HORBERG, L. (1954). Rocky Mountain and Continental Pleistocene deposits in the Waterton region, Alberta. Bull. Geol. Soc. Am., 65: 1093–150.
JÄRNEFORS, B. (1963). Lervarvskronologien och isrecessionen i östra Mellansverige. Sveriges Geol. Undersökn. Årsbok, 57, Ser. C, no. 594.
KARROW, P. F. (1962). Preliminary report on the Pleistocene geology of the Scarborough area. Ont. Dept. Mines, Progr. Rept. 1962-1.
KARROW, P. F., CLARK, J. R., and TERASMAE, J. (1961). The age of Lake Iroquois and Lake Ontario. J. Geol., 69: 659–67.
KAY, G. F. (1931). Classification and duration of the Pleistocene period. Bull. Geol. Soc. Am., 42: 425–66.
KULP, J. L. (1961). Geologic time scale. Science, 133: 1105–14.
LEAKEY, L. S. B., EVERNDEN, J. F., and CURTIS, G. H. (1961). Age of bed 1, Olduvai gorge, Tanganyika. Nature, 191: 478–9.
LEE, T. E. (1957). The antiquity of the Sheguiandah site. Can. Field-Naturalist, 71: 117–37.
LEIGHTON, M. M. (1963). Second preliminary appraisal of the recently proposed classification of the Wisconsin loesses. Mimeographed.
LIBBY, W. F. (1955). Radiocarbon Dating 2nd ed. Chicago: University of Chicago Press.
MacNEISH, R. S. (1951). An archeological reconnaissance in the Northwest Territories. Natl. Mus. Can. Bull., 123: 24–41.
——— (1952). A possible early site in the Thunder Bay District, Ontario. Natl. Mus. Can. Bull., 126: 23–47.
——— (1956). Archeological reconnaissance of the delta of the MacKenzie River and Yukon coast. Natl. Mus. Can. Bull., 142: 46–81.
MASON, RONALD J. (1960). Early man and the age of the Champlain Sea. J. Geol., 68: 366–76.
PREST, V. K. (1957). Pleistocene geology and surficial deposits. In Geology and Economic Minerals of Canada, edited by C. H. Stockwell (Geol. Surv. Can., Econ. Geol. Ser. 1), 4th ed., chap. VII, pp. 443–95.
——— (1961). Soils of Canada. In Soils in Canada, edited by R. F. Legget (Roy. Soc. Can., Spec. Publ. no. 3), pp. 3–21. Toronto: University of Toronto Press.
QUIMBY, G. I. (1960). Indian life in the Upper Great Lakes. Chicago: University of Chicago Press.

Rosholt, J. N., Emiliani, C., Geiss, J., Koczy, F. F., and Wangersky, P. J. (1961). Absolute dating of deep-sea cores by the Pa^{231}/Th^{230} method. J. Geol., *69*: 162–85.

Russell, L. S. (1923). Pleistocene horse teeth from Saskatchewan. J. Paleontol., *17*: 110–14.

—— (1956). Additional occurrences of fossil horse remains in western Canada. Natl. Mus. Can. Bull., *142*: 153–4.

Rutulis, M. (1962). The differentiation of tills in southern Alberta. Unpublished B.Sc. thesis, Department of Geology, University of Western Ontario.

Stalker, A. MacS. (1959). Surficial geology, Fort Macleod, west of fourth meridian, Alberta. Geol. Surv. Can., Map 21-1958.

—— (1960). Surficial geology of the Red Deer – Stettler map-area, Alberta. Geol. Surv. Can., Mem. 306.

—— (1961). Buried valleys in central and southern Alberta. Geol. Surv. Can., Paper 60-32.

—— (1963). Quaternary stratigraphy in southern Alberta. Geol. Surv. Can., Paper 62-34.

Sternberg, G. M. (1930). New records of mastodons and mammoths in Canada. Can. Field Naturalist, *44*: 59–65.

—— (1963). Additional records of mastodons and mammoths in Canada. Natl. Mus. Can., Nat. Hist. Paper no. 19.

Terasmae, J. (1958). Contributions to Canadian palynology. Geol. Surv. Can., Bull. 46.

—— (1960). Contributions to Canadian palynology, no. 2. Geol. Surv. Can., Bull. 56.

—— (1961). Notes on late-Quaternary climatic changes in Canada. Ann. N.Y. Acad. Sci., *95*: 658–75.

Terasmae, J. and Hughes, O. L. (1960). Glacial retreat in the North Bay area, Ontario. Science, *131*: 1444–6.

Tozer, E. T. (1956). Geologic reconnaissance, Prince Patric, Eglinton and Western Melville Islands, Arctic Archipelago, Northwest Territories. Geol. Surv. Can., Paper 55-5.

Wagner, F. J. E. (1959). Palaeoecology of the marine Pleistocene faunas of southwestern British Columbia. Geol. Surv. Can., Bull. 52.

Watt, A. K. (1954). Correlation of the Pleistocene geology as seen in the subway with that of the Toronto region, Canada. Proc. Geol. Assoc. Can., 6 (pt. II): 69–81.

Wilmarth, M. G. (1925). The geologic time classification of the United States Geological Survey compared with other classifications. U.S. Geol. Surv., Bull. 769.

Wormington, H. M. (1957). Ancient man in North America. Denver: Denver Museum, Nat. Hist. Pop. Series, no. 4.

Wright, J. V. (1963). An archaeological survey along the north shore of Lake Superior. Natl. Mus. Can., Anthrop. Paper no. 3.

Lightning Source UK Ltd.
Milton Keynes UK
UKHW030613210722
406167UK00006B/658

9 781487 587031